Blockchain-Enabled Resilience

This book provides a first-of-its-kind approach for using blockchain to enhance resilience in disaster supply chain and logistics management, especially when dealing with dynamic communication, relief operations, prioritization, coordination, and distribution of scarce resources—these are elements of volatility, uncertainty, complexity, and ambiguity (VUCA) describing a dynamic environment that now form the "new norm" for many leaders.

Blockchain-Enabled Resilience: An Integrated Approach for Disaster Supply Chain and Logistics Management analyzes the application of blockchain technology used to enable resilience in a disaster supply chain network. It discusses IoT and DVFS algorithms for developing a network-based simulation and presents advancements in disaster supply chain strategies using smart contacts for collaborations. The book covers how success is based on collaboration, coordination, sovereignty, and equality in distributing resources and offers a theoretical analysis that reveals that enhancing resilience can improve collaboration and communication and can result in more time-efficient processing for disaster supply management.

This book provides a first-of-its-kind approach for managers and policy-makers as well as researchers interested in using blockchain to enhance resilience in disaster supply chains, especially when dealing with dynamic communication, relief operations, prioritization, coordination, and distribution of scarce resources. Practical guidance is provided for managers interested in implementation. A robust research agenda is also provided for those interested in expanding present research.

Blockchain-Enabled Resilience

An Integrated Approach for Disaster Supply Chain and Logistics Management

Polinpapilinho F. Katina
Adrian V. Gheorghe

CRC Press
Taylor & Francis Group
Boca Raton London New York

CRC Press is an imprint of the
Taylor & Francis Group, an **informa** business

First edition published 2023
by CRC Press
6000 Broken Sound Parkway NW, Suite 300, Boca Raton, FL 33487–2742

and by CRC Press
4 Park Square, Milton Park, Abingdon, Oxon, OX14 4RN

CRC Press is an imprint of Taylor & Francis Group, LLC

Library of Congress Cataloging-in-Publication Data
Names: Katina, Polinpapilinho F., author. | Gheorghe, Adrian V., author.
Title: Blockchain-enabled resilience : an integrated approach for disaster supply chain and logistics management / Polinpapilinho F. Katina, Adrian V. Gheorghe.
Description: First Edition. | Boca Raton, FL : Taylor & Francis, 2023. | Includes bibliographical references and index.
Identifiers: LCCN 2022040170 (print) | LCCN 2022040171 (ebook) | ISBN 9781032371504 (hardback) | ISBN 9781032373270 (paperback) | ISBN 9781003336082 (ebook)
Subjects: LCSH: Business logistics. | Humanitarian assistance—Management. Blockchains (Databases)—Industrial applications.
Classification: LCC HD38.5 .K378 2023 (print) | LCC HD38.5 (ebook) | DDC 658.5—dc23/eng/20221212
LC record available at https://lccn.loc.gov/2022040170
LC ebook record available at https://lccn.loc.gov/2022040171

ISBN: 978-1-032-37150-4 (hbk)
ISBN: 978-1-032-37327-0 (pbk)
ISBN: 978-1-003-33608-2 (ebk)

DOI: 10.1201/9781003336082

Typeset in Times
by Apex CoVantage, LLC

Dedications

To my children, Anastasia, Alexandra, and Paul
——Adrian V. Gheorghe

To my sister Rachel, for whom the "resilience" is dedicated
——Polinpapilinho F. Katina

Contents

Foreword

Since the late 1980s, there have been several attempts to create digital cash. In 2009, Satoshi Nakamoto (an anonymous person or group) launched Bitcoin as a peer-to-peer electronic cash. The primary goal was to enable electronic payments to be sent directly from one party to another party without going through an intermediary financial institution. Bitcoin was created by combining several key ideas developed in the fields of cryptography, consensus algorithms, and distributed computer networks. A distributed computer network is a collection of autonomous computers (nodes) that appear to its users as a single coherent computer. Bitcoin is a peer-to-peer type distributed computer network for the realization of an electronic payment system. The electronic payments (or transactions) on the Bitcoin network are validated by miners that are governed by the proof-of-work (PoW) consensus algorithm.

Bitcoin is the first successful electronic payment system that solved the double-spending problem on a peer-to-peer network. Bitcoin inspired the launch of several other alternative coins (referred to as altcoins). Some well-known drawbacks of Bitcoin include high-energy consumption from PoW computations (miners), centralization of mining resources (mining pools), and electronic waste from the disposal of outdated mining hardware. Also, the market value of Bitcoin is speculative in nature and highly volatile, which is uncharacteristic of a stable cash. Despite its drawbacks, bitcoin set the stage for the launch of a new field called blockchain, or more generally speaking, distributed ledger technology (DLT).

In 2015, programmable blockchain/DLT projects such as Ethereum, Hyperledger Fabric, and R3 Corda were launched. Over the last decade, the field of blockchain/DLT has evolved to encompass a collection of distributed computer network architectures implementing various data structures and consensus algorithms. Blockchains and DLTs can be categorized based on the type of ledger and participation of validators:

- *Public-Permissionless Type:* The ledger is visible to the public, and anyone can join the network. Also, the validators can be anyone on the network. These are fully decentralized architectures (e.g., Bitcoin).
- *Public-Permissioned Type:* The ledger is visible to the public. However, the validators are selected by a governing body or a consensus algorithm. Typically, these are semi-centralized architectures (e.g., Binance Smart Chain).
- *Private Type:* The ledger is private and visible to only the members of the network. A governing body selects the validators. These are centralized architectures (e.g., Hyperledger Fabric).

Cryptocurrencies are primarily based on public-type blockchains. Hundreds and thousands of innovative startup ideas, projects, and companies are being built and experimented on public blockchains on topics such as stablecoin, decentralized finance (DeFi), non-fungible tokens (NFTs), decentralized autonomous organizations (DAOs), games, and the metaverse. Still, in its nascent stage, the field has become notorious for hacks, scams, and bugs in the code, where millions or even billions have been lost in a matter of seconds or minutes by startups. As the technology matures, the hope is that many of these issues will be addressed to enable Web 3.0.

Enterprise and government applications are typically built on private-type blockchains. Some enterprise case studies include the food supply chain, the pharmaceutical supply chain, the metal mining supply chain, invoice management, the insurance industry, and healthcare claims. Some government use cases of blockchain technology include identity, voting, national digital currency, land registry, budgeting, taxation, border and customs control, public procurement, and e-residency.

All mainstream industries such as agriculture, food, shipping, manufacturing, pharmaceuticals, etc. employ specialized supply chains. The current digital infrastructure supporting them remains fragmented, inefficient, and expensive. Typically, participants in the current supply chain carry out transactions on their own ledgers. Thus, facts are scattered on multiple ledgers, leaving room for errors, fraud, and inefficiencies. Using blockchain technology, supply chain participants from within one organization or multiple organizations can share a distributed ledger that contains a record of the history of transactions with internal consistency. Once a transaction is recorded in the ledger, it is immutable and thus cannot be altered. Thus, the ledger provides greater end-to-end visibility into the provenance of data among the participants.

Blockchain-based supply chains can help organizations manage the workflow of data and the movement of data both from internal and external sources (e.g., finance and banking industry). Blockchain-based supply chains can also improve the accuracy, efficiency, and transparency of the various contracts within the workflow of a given supply chain management system (e.g., procurement, shipping industry). Moreover, blockchain-based supply chains can track the movement and distribution of goods in combination with data from the Internet of Things (IoT). This will provide better transparency for detecting and preventing the circulation of counterfeit goods (e.g., medicine, manufacturing parts, luxury products).

The COVID-19 pandemic exposed the inefficiencies and deficiencies of supply chains in various sectors. Governments all over the world were unprepared for such a disaster on a global scale. Notably, no supply chain solution was available for fast and efficient delivery of vaccines under refrigeration to various vaccination centers. This book is very timely and focuses on the potential of blockchain for resilience in supply chain and logistics networks. The state-of-the-art topics covered in this book are relevant to academic researchers, industry professionals, and policy-makers that are exploring emerging technologies for disaster management.

San Juan, Puerto Rico   **Ramesh Ramadoss, PhD**
July 2022   Chair, IEEE Blockchain Initiative

Preface

Humans are living in a moment in time when, perhaps due to technological innovation, they praise themselves with the belief of being well informed, intelligent, wiser, and capable of making even better judgments—and yet incapable of explaining why doomsday fears loom large. Volcanoes continue to kill plants and animals, supervolcanoes threaten whole species with extinction, diseases still run rampant, and nuclear weapons remain one of the gravest and most lethal threats to humans.

Since World War II, remarkable technological advances have affected agriculture, economy, education, finance, medicine, public health, telecommunications, transportation, and water—and these have affected longevity and living standards. For example, a United Nations report suggests that "globally, a person aged 65 years in 2015–2020 could expect to live, on average, an additional 17 years. By 2045–2050, that figure will have increased to 19 years. Between 2015–2020 and 2045–2050, life expectancy at age 65 is projected to increase in all countries" (United Nations, 2019, p. 1). Unfortunately, technology has also been at the forefront of crime and violence (EUROPOL, 2017). It is now possible to commit a crime in one part of the world while living in another. For example, Bitfinex, a Bitcoin exchange based in Hong Kong, was hacked in 2015, losing about $400,000. In 2016, the same entity lost about $73 million more when it was stolen from customers' accounts by hackers in the United States (Nakamura, 2017; Katina et al., 2019). Also, on May 7, 2021, Colonial Pipeline, an American oil pipeline system that originates in Houston (Texas, USA), suffered a ransomware cyberattack that impacted computerized equipment managing the pipeline. The company halted all pipeline operations and paid over $4 million in Bitcoins to restore services (Turton et al., 2021).

Clearly, there is an unforgiving imbalance between good and bad associated with technology. *But is technology the problem? Or are humans the problem?* This book does not try to answer this question. However, we offer the following as a means to ponder on where we go from here: First, individuals that were once considered humane and trustworthy are becoming more and more unreliable, inconsistent, incompetent, and downright corrupt with no moral sense. Second, and on technology, we echo complexity scientist Dr. Samuel Arbesman's (2016, p. 2) words: "Our technologies—from websites and trading systems to urban infrastructure, scientific models, and even the supply chains and logistics that power large businesses—have become hopelessly interconnected and overcomplicated, such that in many cases even those who build and maintain them on the daily basis can't fully understand them any longer."

The aforementioned hints at complexity—the identifying mark of problems facing practitioners in modern society, permissive and intractable given the apparent ineffectiveness of the responses the present landscape. However, complexity is only one element of elements that one must be content with in navigating the 21st-century landscape. In addition to dealing with complexity, which describes how various factors interact without any clear cause and effect patterns (Bennis's (2001), research suggests dealing with volatility (the speed of change currently encountered), uncertainty (the lack of predictability and the likelihood of unexpected events), and ambiguity (the difficulty in interpreting conditions and current reality). *Volatility, uncertainty, complexity,* and *ambiguity,* or VUCA, represents the operational environment from which systems and organizations of the 21st century must operate. These conditions continue to proliferate into all aspects of human endeavor and the systems designed to orchestrate those endeavors. They are not the privilege, or torment, of any particular field or sector (e.g., agriculture, economy, education, energy, finance, medicine, public health, telecommunications, transportation, or services), as none are immune to the effects of VUCA.

After the Colonial Pipeline ransomware attack, US President Joe Biden declared a state of emergency on May 9, allowing for temporary suspension of limits on the amount of petroleum products transported domestically within the U.S mainland (Russon, 2021). Thus, it's not for a lack (or implementation) of policy that problems persist; rather, problems persist in spite of policy. It appears that "good will" is no match for "messes" Ackoff (1981) and "wicked problems" (Rittel & Webber, 1973). A mess is an interrelated set of poorly formulated problems not easily understood or resolved. A wicked problem is intractable with contemporary thinking, decision, action, and interpretation. Messes and wicked problems reside in a landscape requiring a "holistic" approach (Smuts, 1926). And for this, we humbly suggest emerging research on Complex System Governance (Keating et al., 2022). However, the present volume is an attempt dedicated to redeeming technology as a necessary and useful tool despite the foregone challenges.

As already suggested, VUCA describes the constant, unpredictable change situations that now form the "new norm" for many leaders in many problem domains; it is no exception regarding disaster management. A disaster and its area are dynamic. The blockage can be in supply distribution coming from different sources, including damaged infrastructure affecting the availability and accuracy of information. Success in the disaster supply chain and logistics network is based on collaboration and coordination in distributing resources. Managers in such situations are expected to make good decisions (transparent, trusted, and traceable), and as such, they must be able to balance the uncertainty in the decision-making process. A promising technological innovation of blockchain technology promises real-time, secure, and transparent information exchange and automation via smart contracts. This study analyzes the application of blockchain technology as a means to enable resilience in a disaster supply chain and logistics management. Resilience in a disaster supply network involves the ability of individuals, communities, organizations, and states to adapt to and recover from hazards, shocks, or stresses without compromising long-term prospects for development using innovative means. In this research, the innovative means of blockchain technology and the dynamic voltage frequency scaling (DVFS) algorithm are the basis for developing a network-based simulation to enhance the resiliency of the disaster supply network.

As such, this book promotes emerging technologies to enhance resilience in disaster management, especially in the face of increased disasters and the associated uncertainties and complexities. This research should attract policy-makers' attention, especially since they are ultimately responsible for disaster relief decision-making. However, researchers (as well as students) should pay close attention to how this book cannibalizes and uses several well-established methods, tools, techniques, and topics (e.g., resilience and merging technologies) for research.

With this audience in mind, seven chapters have been developed. **Chapter I** discusses the fundamentals of supply chain management, system resilience, and blockchain technology, along with a need for blockchain-enabled resilience in supply chain and logistics networks, especially in disaster-stricken areas. This chapter section sets the stage for the remainder of the book.

Chapter II articulates the many faces of disaster management and the need for a flexible and multifaceted strategy for disaster relief. This chapter suggests that dealing with disaster is a complex issue involving social-technical aspects that do not fit precisely within the scope of traditional management. Therefore, there is a need to rely on the diffusion of many norms, state regulations and self-regulation, mechanisms of the market, and other processes (negotiation, participation, and engagement) that facilitate collective decision-making and activities.

Chapter III articulates theories of disaster management, theories of disaster supply chain, and the role of modeling and simulation in disaster management. The role of emerging technologies (e.g., blockchain technology, IoT, and 6G) is then discussed in the context of disaster management. DVFS is proposed as a means to sort processes embedded in the server for analyzing the complexity of disaster management complexity. The chapter concludes with a systematic review of topics, challenges, and blockchain contributions to the supply chain knowledge base—a basis for the present research.

Chapter IV details the methodology for the blockchain-enabled model for disaster supply management, including research design, the structure of the model, and the framework for implementation. The role of modeling and simulation in drawing attention to physical experimentation, including disaster management, is presented. The simulation modeling approach is efficient since it avoids actual physical experimentation, which can be costly, time-consuming, and deadly in disaster situations. The chapter then indexes the suggested model and explains how blockchain technology can be used with IoT to address disaster supply chain and logistics management communication challenges.

Chapter V presents the outcomes of 100 iterations modeling the DVFS model using a data stream. Critical requirements in disaster response operations are articulated. The proposed algorithm and a simulation design that ensures repeatability are then provided. The simulation is then executed under four different conditions: a smart-contracts-enabled simulation for hyperconnected logistics (SCESHL) model (serving as a baseline), the DVFS algorithm applied to the SCESHL model, a proposed blockchain-enabled DVFS-based model, and a proposed blockchain-enabled DVFS-based model integrated with multi-agent systems. The levels of improvement in the disaster supply chain network are then discussed along six specifically selected parameters: energy consumption, number of actions, successful migrations, system errors, system throughput, and system delays.

Chapter VI focuses on the developed model disaster supply chain and logistics management implications and beyond. Technical innovations and breakthroughs happening now are outlined. The findings associated with the application of an integrated blockchain DVFS-based model for enhancing the performance of a disaster supply network are provided in the context of present research trends (and limitations) along with implications. The chapter concludes with the possible paths that extend present research along resilience indicators, practical tools, and case applications.

The final chapter is **Chapter VII**, and it provides an initial research agenda on blockchain-enabled resilience for disaster supply chain and logistics management. It provides a guiding plan intended to stretch the limitations of the present research to include critical issues of knowledge claims—ontology, epistemology, methodology, and the nature of human beings. The chapter includes a framework purposefully developed for interrelated lines of inquiry of blockchain-enabled resilience for disaster supply chain and logistics management along philosophical, theoretical, axiological, methodological, axiomatic, and applications.

The book also includes glossary listing terms often used in disaster preparedness and their definitions. In general, explanations of concepts are relevant to the current research. However, the reader might also reference the listed concepts and their meaning elsewhere.

Polinpapilinho F. Katina
Spartanburg, South Carolina, USA

Adrian V. Gheorghe
Zürich, Switzerland

REFERENCES

Ackoff, R. L. (1981). The art and science of mess management. *Interfaces*, *11*(1), 20–26.

Arbesman, S. (2016). *Overcomplicated: Technology at the limits of comprehension*. Current.

Bennis, W. (2001). Leading in unnerving times. *MIT Sloan Management Review*, *42*(2), 97–102.

EUROPOL. (2017). Crime in the age of technology (EDOC# 924156 v7). *The European Union Agency for Law Enforcement Cooperation*. www.cepol.europa.eu/sites/default/files/924156-v7-Crime_in_the_age_of_technology_.pdf

Katina, P. F., Keating, C. B., Sisti, J. A., & Gheorghe, A. V. (2019). Blockchain governance. *International Journal of Critical Infrastructures*, *15*(2), 121. https://doi.org/10.1504/IJCIS.2019.098835

Keating, C. B., Katina, P. F., Chesterman, C. W., & Pyne, J. C. (Eds.). (2022). *Complex system governance: Theory and practice*. Springer International Publishing. https://link.springer.com/book/10.1007/978-3-030-93852-9

Nakamura, Y. (2017, May 21). *Bitfinex comes back from $69 million bitcoin heist*. SFGate. www.sfgate.com/business/article/Bitfinex-comes-back-from-69-million-bitcoin-heist-11161585.php

Rittel, H. W. J., & Webber, M. M. (1973). Dilemmas in a general theory of planning. *Policy Sciences*, *4*(2), 155–169. https://doi.org/10.1007/BF01405730

Russon, M.-A. (2021, May 10). US fuel pipeline hackers "didn't mean to create problems." *BBC News*. www.bbc.com/news/business-57050690

Smuts, J. (1926). *Holism and evolution*. Greenwood Press.

Turton, W., Riley, M., & Jacobs, J. (2021, May 13). Colonial ppipeline paid hackers nearly $5 million in ransom. *Bloomberg.Com*. www.bloomberg.com/news/articles/2021-05-13/colonial-pipeline-paid-hackers-nearly-5-million-in-ransom

United Nations. (2019). *World population ageing 2019: Highlights* (ST/ESA/SER.A/430). Department of Economic and Social Affairs. www.un.org/en/development/desa/population/publications/pdf/ageing/WorldPopulationAgeing2019-Highlights.pdf

Acknowledgment

One aspect of the academic citation apparatus is acknowledging the relevance of the works of others to the topic of discussion. This we have done. However, this apparatus fails to capture the influences of many that were involved in current efforts. With this in mind, the editors wish to acknowledge different people and organizations involved in discussions of this research.

First, we want to recognize the stimulating comments and criticisms of Prof. Dr. Enrico Zio of Politecnico di Milano (Milan, Italy) and CentraleSupélec (Paris, France). Over the years, several faculty members helped by reviewing changes and providing feedback. Many industrial practitioners also assisted us by commenting on draft chapters. We also wish to acknowledge several students, many of whom we proudly now call colleagues, in classes where we tested teaching/research materials: Ange-Lionel Toba, B. Tyler McDaniel, Casey Cash, Dean Vanadore, Derrick Talley, Farinaz Sabz AliPour, Fredderick Coleman, Herbert James, James Bobo, Jeffery Sullivan, Joel Pridmore, John Kiehl, Jonathan Linder, Justin Bennefield, Logan Caldwell, Michael Milas, Nima Shahriari, Omer Keskin, Omer Poyraz, Richard Gallego, Roberto Olvera, Sean Bartels, Sujatha Alla, Unal Tatar, and Ying Thaviphoke.

Also, beyond the authors' toiling, this book is a measurable expression of their intense intellectual interaction and cross-fertilization of ideas with several distinguished colleagues and partners in mind from the Academia and academic entities of many denominations. Most significantly, our gratitude goes to the following: Adolf J. Dörig, Dörig + Partner AG (Salzburg, Austria); Cornel Vintila, AuraChain (Bucharest, Romania); Dr. Dan V. Vamanu (Horia Hulubei National Institute for R&D in Physics and Nuclear Engineering (Bucharest, Romania); Dr. Ona Egbue, University of South Carolina Upstate (Spartanburg, SC, USA); Dr. Roland Pulfer (Action4Value, Kirchseelte, Germany); Jürg Birchmeier, Zürich Insurance Company Ltd (Zürich, Switzerland); Laura Manciu, AuraChain (Bucharest, Romania); Marcelo Masera, Joint Research Centre (Petten, The Netherlands); Paul Niculescu-Mizil Gheorghe (ICI Bucharest, Bucharest, Romania); Prof. Dr. Charles Keating, Old Dominion University (Norfolk, VA, USA); Prof. Dr. Ioannis Papazoglou, National Centre for Scientific Research 'DEMOKRITOS' (Athens, Greece); Prof. Dr. Radu Cornel, University Politehnica of Bucharest (Bucharest, Romania); Prof. Dr. Wolfgang Kröger, ETH Zürich (Zürich, Switzerland); Tim Ellis, University of South Carolina Upstate (Spartanburg, SC, USA).

The authors are also grateful to graduate students and young researchers in the Department of Informatics and Engineering Systems, University of South Carolina Upstate, USA, and the Department of Engineering Management and Systems Engineering, Old Dominion University, USA.

For help with preparing the manuscript, we are thankful to Allison Shatkin, Brian Romer, and Paul Niculescu. We want to particularly thank our executive editor Cindy R. Carelli for her encouragement throughout the project.

Our sincere apologies to everyone else who ought to have been remembered here.

Polinpapilinho F. Katina
Spartanburg, South Carolina, USA

Adrian V. Gheorghe
Zürich, Switzerland

About the Authors

Polinpapilinho F. Katina is an assistant professor in the Department of Informatics and Engineering Systems at the University of South Carolina Upstate (Spartanburg, South Carolina, USA). He has served in various capacities at the National Centers for System of Systems Engineering (Norfolk, Virginia, USA); Old Dominion University (Norfolk, Virginia, USA); Politecnico di Milano (Milan, Italy); the University of Alabama in Huntsville (Huntsville, Alabama, USA); and Syracuse University (Syracuse, New York, USA). Dr. Katina holds a BS in Engineering Technology, an MEng in Systems Engineering, and a PhD in Engineering Management and Systems Engineering (Old Dominion University, Norfolk, Virginia, USA). He received additional training at the Politecnico di Milano (Milan, Italy).

He focuses on teaching and research in the areas of complex system governance, critical infrastructure systems, decision-making and analysis, emerging technologies (e.g., IoT), energy systems (smart grids), engineering management, infranomics, manufacturing systems, system of systems, systems engineering, systems pathology, systems theory, and systems thinking. He has demonstrable experience leading large-scale research projects and has achieved many established research outcomes.

His profile includes nearly 200 scholarly outputs of peer-reviewed journal articles, conference papers, book chapters, and technical reports. He has also co-authored six books (three monographs and three edited). Dr. Katina is a reviewer for several journals and serves on the editorial board for *MDPI*. He is an editor for *John Wiley & Sons/Hindawi* and *Inderscience*. He is a senior member of IEEE and ASEM. He is a recipient of several awards, including *Excellence in Teaching and Advising* (University of South Carolina Upstate), top *1% for the 2018 Publons Global Peer Review Awards*, and *2020 IAA Social Sciences Book Award* (IAA: International Academy of Astronautics).

Adrian V. Gheorghe is an engineering management and systems engineering professor and the Batten Endowed chair on system of systems engineering with the Department of Engineering Management and Systems Engineering at Old Dominion University (Norfolk, Virginia, USA). Prof. Dr. Gheorghe holds an MSc in Electrical Engineering (Bucharest Polytechnic Institute, Bucharest, Romania); a PhD in Systems Science/Systems Engineering (City University, London, U.K.); an MBA from the Academy of Economic Studies (Bucharest, Romania); and an MSc Engineering Economics (Bucharest Polytechnic Institute, Bucharest, Romania).

Dr. Gheorghe is a senior scientist with the European Institute for Risk and Communication Management (Bucharest, Romania) and vice president of the

World Security Forum (Langenthal, Switzerland). He has worked with different organizations, including Battelle Memorial Institute (Columbus, Ohio); Beijing Normal University (Beijing, China), International Atomic Energy Agency (Vienna, Austria); International Institute for Applied Systems Analysis (Laxenburg, Austria); Joint Research Centre of the European Commission (Ispra, Italy); Riso National Laboratory (Roskilde, Denmark); Stanford University (Stanford, California); Swiss Federal Institute of Technology (Zürich, Switzerland); and United Nations University (Tokyo, Japan).

His profile includes more than 400 scholarly outputs of peer-reviewed journal articles, conference papers, book chapters, and technical reports. He has published several books, including *Emergency Planning Knowledge* (VdF Verlag, 1996); *Integrated Risk and Vulnerability Management Assisted by Decision Support Systems: Relevance and Impact on Governance* (Springer, 2005); *Critical Infrastructures at Risk: Securing the European Electric Power System* (Springer, 2006); *Critical Infrastructures: Risk and Vulnerability Assessment in Transportation of Dangerous Goods—Transportation by Road and Rail* (Springer, 2016); and *Critical Space Infrastructures: Risk, Resilience and Complexity* (Springer, 2019). Dr. Gheorghe is an editor of several journals, including *International Journal of Critical Infrastructures* and *International Journal of System of Systems Engineering*. He is a reviewer for several journals, including the *International Journal of Technology Management*. He has served as a guest editor for several journals, including *International Journal of Environment and Pollution, International Journal of Global Energy Issues*, and *International Journal of Technology Management*.

1 Supply Chain Management in the Age of Complexity

1.1 SUPPLY CHAIN MANAGEMENT

Organizations in the 21st century increasingly need to rely on effective supply chains and networks to compete in the globally interconnected world. In recent decades, globalization, outsourcing, and information technology have enabled many organizations, such as Dell and Hewlett Packard, to successfully operate collaborative supply networks in which each specialized business partner focuses on only a few key strategic activities. Irrespective of system or network, society faces what appears to be an intractable problem domain for systems essential to societal well-being (e.g., health, food, transportation, energy, security). Several dominant characteristics might capture the operational landscape. Keating and Katina (2019) suggest dominant characteristics of complexity, context, ambiguity, and holism:

- *Complexity:* involves exponentially increasing amount, availability, veracity, and accessibility of information coupled with the increasingly large number of richly interconnected elements. Involves incomplete, fallible, and dynamically evolving system knowledge and high levels of uncertainty beyond current capabilities to structure, order, and reasonably couple decisions, actions, and consequences. Complexity also encompasses the emergence of behavior, performance, and consequences that cannot be known or predicted before their occurrence
- *Contextual Dominance:* involves unique circumstances, factors, patterns, and conditions within which a system is embedded, influencing the system andconstraining/enablingsystemperformance.Itaffectsdecisions,actions,and interpretations made with respect to the system. Also, it highlights multiple stakeholders with different worldviews (convergent/divergent), objectives, and influence patterns.
- *Ambiguity:* involves instabilities in understanding system structure, behavior, or performance. There is a potential lack of clarity in system identity/purpose, boundary conditions, delineation of system constituents, or understanding of a system within its context.
- *Holism:* involves technical/technology aspects of a system—the entire influencing spectrum of human, social, organizational, managerial, policy, political, and information aspects is central to a more complete (holistic) system view. Behavior, along with properties and performance, is a function of interactions that cannot be revealed by understanding the constituents.

DOI: 10.1201/9781003336082-1

Supply chains are not immune to the aforementioned landscape. For example, during the early months of COVID-19, governmental organizations failed to keep pace with the spread of the virus. The devastating COVID-19 crisis turned many sectors upside down in the US, including the country's storied logistics industry. Some organizations were able to develop foreign supply chains to import much-needed medical supplies quickly (Atkins, 2020; García-Herrero, 2020). Supply chain management (ASCM, 2022; Cornell Engineering, 2022), which can be defined as the design, planning, execution, control, and monitoring of supply chain activities to create net value, competitive infrastructure, leveraging logistics, synchronizing supply with demand, and measuring performance globally, proved critical in attempts to holistically navigate the complexity, contextual, and ambiguity of fighting against COVID-19.

The supply chain management practice draws heavily on industrial engineering, systems engineering, operations management, logistics, procurement, information technology, and marketing (Kozlenkova et al., 2015). It strives to be an integrative, multidisciplinary, and multimethod approach to crises (Sanders & Wagner, 2011). Moreover, there is also a trend toward sustainability and risk management in supply chain management (Lam, 2018), with some suggesting a need to include people dimension, ethical issues, internal integration, transparency/visibility, and human capital/talent management within the SCM research agenda (Wieland et al., 2016).

Moreover, the operations of many organizations are supported by the supply chain. For example, the US Department of Defense (DoD) operations' capabilities are supported by complex supply chains (Alberts et al., 2017). The complexities arising from various factors, such as changes in customer expectations, multiple market channels, and international markets, create significant challenges throughout the supply networks. New models are required to support future supply chain management (Ivanov et al., 2019). Because readers might only be interested in some aspects of this book, a flow diagram is provided in Figure 1.1.

1.1.1 DIMENSIONS OF SUPPLY CHAIN MANAGEMENT

A supply chain is a set of three or more entities directly involved in the upstream and downstream flows of products, services, finances, or information from a source to a customer (Mentzer et al., 2001). A supply chain is a network of multiple businesses and relationships considered a complex system due to having multiple levels, numerous facilities at each level, and being dispersed over a large geographical location (Beamon, 1999; Lambert & Cooper, 2000). The critical objectives of the supply chain are determined by Kshetri (2018) as cost, quality, speed, dependability, risk reduction, sustainability, and flexibility. Coordination among the supply chain organizations on tactical, operational, and strategic levels is essential for an effective supply chain (Ludema, 2002).

Supply chain management represents a novel way of business management and relationships for the chain members' total business process excellence (Lambert & Cooper, 2000). If one takes the definition of SCM as suggested by Mentzer et al. (2001)—"the systemic, strategic coordination of the traditional business functions and the tactics across these business functions within a particular company and across businesses within the supply chain, to improve the long-term performance

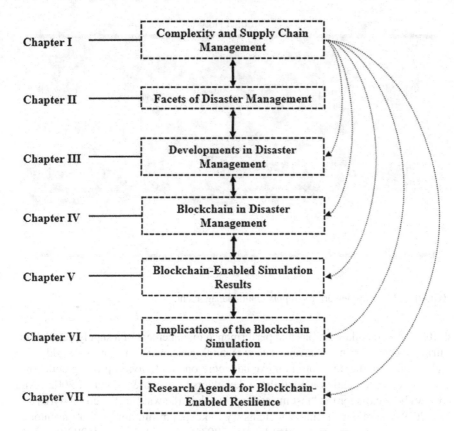

FIGURE 1.1 Book flowchart indicating relationships and research themes.

of the individual companies and the supply chain as a whole" (Mentzer et al., 2001, p. 18)—then supply chain management can be grouped into operation, design, and strategy categories as suggested by Huan et al. (2004). Moreover, Yoo and Won (2018) suggest an SCM framework inclusive of a bidirectional flow of products and information, as indicated in Figure 1.2.

Interaction of different groups of a system with each other and information sharing play a critical role in the business's success. A standard representation for information sharing can be obtained via a supply chain framework that facilitates communication among different parties (Sharawi et al., 2006). A supply chain is ultimately a human activity and a multidisciplinary approach where systems engineering would be an appropriate foundation to tackle the challenges in cross-organizational supply chains (Haskins, 2006).

The evolution of supply chain management is rooted in the 1960s when the focus was on minimizing production costs. In the 1970s, the focus shifted to material requirement planning. In the 1980s, the concept of "just in time" was developed, which offers low-cost, high-quality, and reliable products. In the 1990s, the focus was more on integrations among different involved parties in a supply network. The focus

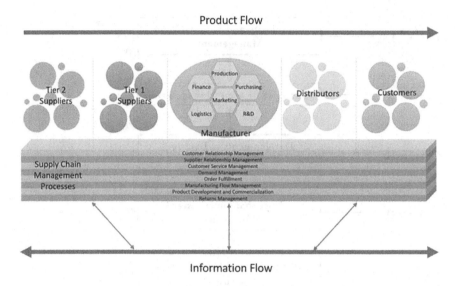

FIGURE 1.2   A framework for supply chain management.

shifted more to collaborations and performance management of a supply network for a firm's success. With the advent of new technologies and the use of the World Wide Web (Web 2.0), the concentration extends more on creativity, improving collaboration, and communication among the stakeholders (Fawcett & Magnan, 2002). And now, we hear calls for the "just in case" approach in the wake of the coronavirus disease 2019, where our inventory systems try to keep a minimum level of inventories just in case of emergencies (Griffy-Brown, 2003; Schwab & Malleret, 2020).

The supply chain management framework relies on the distinction between function-oriented and organization-oriented focus. Function-oriented focus is on purchase and supply, logistics, transport, marketing, and business management. Organizational-oriented focus is on industrial organization, supply chain configuration, transaction costs, and system dynamics. The latter emphasizes the networked nature of supply chains and the processes that shape them (Park et al., 2013). Hence, supply chain management success is related to synchronizing various activities within the network (Golicic & Davis, 2012). Performance measurement is an essential strategic tool to achieve the acquired objectives and fulfill the company's mission (Fawcett & Magnan, 2002). There are different aspects of performance measurement systems in the supply chains. For example, (Beamon, 1999) suggests resources, output, and flexibility while Kshetri (2018) suggests cost, time and speed, dependability, risk reduction, sustainability quality, and flexibility—measures that are widely accepted (Fawcett & Magnan, 2002; Kamble & Gunasekaran, 2020; Neely et al., 1995).

The extended enterprise view has recognized that the companies compete on coordinated supply chains or a network of companies (Haskins, 2006). The critical elements for the success of supply chain integration are the development and integration of people and technological resources along with the coordinated management of materials, information, and financial flows (Haskins, 2006). The roles of different actors involved and the supply chain management are dependent on the company and the industry. Fawcett and Magnan (2002) provided a systematic literature review that helps understand the main activities and a complete picture of the supply chain. Their findings show that advanced techniques such as simulation, artificial neural networks, and fuzzy logic have been applied for decision-making in the supply chain (Fawcett & Magnan, 2002). Park et al. (2013) argue that the literature has developed a hierarchy of relationships on the degree to which a lead firm controls the entire supply chain, which is dependent on information requirements, the capacity for product differentiation, the inherent complexity of the product, the degree of durability needed in a relationship, and the market power.

Furthermore, Fawcett and Magnan (2002) suggest that challenges for supply chain management include cultural development (based on ongoing and shared learning and continuous improvement) and the emergence of the network organization (which has a complex web of linkages that needs coordination and management). The emergence of the network organization can add more complications, such as a lack of openness (and opportunistic behavior), conflict over autonomy and accountability, culture and procedures, lack of common purpose, multiple (and hidden) goals, overdependence, and power imbalances.

1.1.2 CHARACTERISTICS OF SUPPLY CHAIN MANAGEMENT

The focus of the present work is supply chain and logistics challenges arising from the communication and collaboration within the network. The term "supply network" has been used since the 1990s to describe the dynamic, interconnected, complex, interdependent network of suppliers, manufacturing facilities, and other relevant organizations (Bales et al., 2004). A supply network comprises the member companies and the links between them. Three primary aspects of a network structure suggested by Lambert and Cooper (2000) are the members, the structural dimensions of the network, and the different types of process links across the supply chain.

Characteristics of a supply network adopted via literature include density and network size (Basole et al., 2016; Craighead et al., 2007); complexity (Craighead et al., 2007; Saberi et al., 2019; Wang et al., 2018); risk level (Basole et al., 2016); mutual trust (Terzi & Cavalieri, 2004; Wang et al., 2018); information sharing (Terzi & Cavalieri, 2004; Wang et al., 2018); and collaboration (Terzi & Cavalieri, 2004; Wang et al., 2018). The integration and operation of the supply network require continuous information flow to create the most significant flow for the product. The critical point for having effective supply network management is controlling uncertainty in customer demand, manufacturing processes, and supplier performance (Lambert & Cooper, 2000).

This section outlines a brief history and the importance of effective supply chains and logistics networks. Despite developments in supply chains and logistics networks, challenges remain, especially with the increasing complexity of supply and logistics networks, coupled with operational context, ambiguity, and the need to be a holistic approach when dealing with emergencies. At this confluence, the remainder of this book attempts to contribute to the practice of supply chain management by developing and applying a blockchain-enabled resilience approach for supply chain and logistics management while focusing on disaster relief. The remainder of this chapter addresses resilience and blockchain while focusing on disaster relief.

1.2   SYSTEM RESILIENCE

Suppose one takes the view that resilience is the ability to prepare for and adapt to changing conditions and withstand and recover rapidly from disruptions and that it includes the ability to withstand and recover from deliberate attacks, accidents, or naturally occurring threats or incidents. Then, a resilient supply chain and logistics network must have such an ability. However, the preceding section suggests that the operating landscape of supply chains and logistics is increasingly complex and ambiguous. In such an environment, ongoing and shared learning, continuous improvement, and consideration of emergent networks are necessary. However, the emergent networks add more difficulties (e.g., lack of common purpose, multiple and hidden goals, power imbalances, culture and procedures, conflict over autonomy and accountability, overdependence, and lack of openness and opportunistic behavior) in effectively managing the supply chains and logistics. And yet, such challenges create the right environment for additional risks. Moreover, the risk does not exist in isolation—it thrives on the accompanying vulnerability, fragility, and perception: all these should somehow be addressed to make supply chains and logistics more resilient. Therefore, developing a resilient supply chain and logistics network must also account for risk and the related topic. These relationships are the subject of the following sections.

1.2.1   SYSTEM RISK

The term "risk" enjoys many definitions, with no one description unanimously accepted. It has been present, used, and debated for years (Holton, 2004; Knight, 1921). Risk has been defined as the probability of occurrence of an event and the magnitude of the expected consequences (ASCE, 2009). However, it is also associated with uncertainty. In the system life cycle, risk is also associated with uncertainty and opportunities related to cost, schedule, and performance (INCOSE, 2011). In decision-making, risk is associated with probabilities of unknown outcomes and uncertainty (Gibson et al., 2007). Risk has also been defined as "the potential that something will go wrong as a result of one or a series of events" (Blanchard, 2008, p. 344) and is equated to "a probability event" (Garvey, 2009, p. 33).

Moreover, Hill (2012) theorizes that risk can also be viewed as subjective since it can be a mental construct. Arguably, supply chains and logistics networks are *exposed* to different types of risks since they do not operate in isolation in relation to internal and external interfaces. Furthermore, it is reasonable to assume that supply chains and logistics networks are always *under threat* from naturally occurring events such as flooding, drought, pandemics, and malicious attacks. In assessing risk for such systems, an objective could be a simple risk classification (e.g., acceptable/unacceptable) or allocating scarce resources to manage the risk.

Furthermore, it stands to reason that some domains (e.g., nuclear, aerospace, oil, rail, and defense) will have well-defined risk assessment methods due to the long-standing availability of statistical data rooted in historical accounts and consolidated safety culture. And yet others—for example, cyber-physical systems—will need new methods and tools for assessing emerging threats (e.g., cyberattacks). Despite this rich and sometimes confusing conceptual landscape, central to risk remains a construct defined in probability and consequence, visualized with a *risk matrix* conducive to a mitigative policy based on a corrective intervention *as low as reasonably achievable* (ALARA). Nevertheless, developing a resilient supply chain and logistics network requires consideration of risk in some form or another.

1.2.2 System Vulnerability

Similar to risk, the concept of vulnerability has many definitions, with no one unanimously accepted definition (Katina et al., 2014). In fact, vulnerability was long considered as being closely similar to risk, if only with a broader interpretation. However, Song (2005) notes that some authors clearly distinguish between vulnerability and risk. For example, Turner et al. (2003) depict vulnerability as a degree to which a system, subsystem, or system component is likely to experience harm due to exposure to a hazard, either a perturbation or a stress/stressor. Einarsson and Rausand (1998) and Holmgren et al. (2001) vulnerability is defined as the properties of a system that may weaken or limit its ability to survive and perform its mission in the presence of threats that originate both within and outside the system boundaries. Song's (2005) research establishes a critical difference between vulnerability and the degree of vulnerability: vulnerability is the susceptibility and resilience/survivability of the community/system and its environment to hazards, where susceptibility comprises two aspects: exposure and sensitivity. Survivability mainly comprises robustness, reliability, redundancy, and adaptation (Song, 2005). In this case, the degree of vulnerability is the numerical index of the vulnerability based on different criteria, usually ranging from 0% to 100%. Aven's research (2011) evokes a common definition: "manifestation of the inherent states of the system that can be subjected to a natural hazard or be exploited to adversely affect that system" (Aven, 2011, p. 515).

Regardless of diverging perspectives on vulnerability definition, there is consensus on the need to consider vulnerability in system assessment. The International Risk Governance Council stipulates that vulnerability is a viable area of research, especially in critical infrastructure systems, where disaster management belongs to the emergency services sector. This might be attributed to "basic weaknesses, such as over-complexity and traded-off security factors, and [the fact that such systems] face multiple threats, including exposure to natural hazards and malicious attacks" (IRGC, 2007, p. 4). The issue of vulnerability is essential in supply chains and logistics since such networks operate in the open; they are exposed to different elements.

A clear demarcation between vulnerability and risk can also be seen in the assessment. Song (2005) suggests that risk assessment involves selecting particular stress (hazard) of concern and identifying consequences on the system. Vulnerability assessment should look at a particular system (or component) and examine how it can be affected by various stressors. Obviously, such an analysis will involve the identification of means to reduce vulnerability (Tokgoz & Gheorghe, 2013). In Song's clear-cut words, vulnerability describes the "inherent characteristics of a system that create the potential for harm but are independent of the risk of occurrence of any particular hazard" (Song, 2005, p. 19). Apart from the manner of assessment, differences can also be pointed at in the scope of the analysis and emphasis (Einarsson & Rausand, 1998; Song, 2005).

Again, like with risk, different models exist for dealing with vulnerability. Further discussions on these can be found elsewhere (Vamanu et al., 2016). In the present text, the *quantitative vulnerability assessment* (QVA) is central. QVA results from a warranted analogy with *quantitative risk assessment* (QRA)—a term coined in the closing decade of the past century and that has made quite a career in the community of risk and safety managers worldwide. Like its risk-related counterpart, QVA is about expressing its object—vulnerability—in numbers in a scientifically defendable and practically meaningful way.

In the QVA approach, vulnerability is described as "a predictive quantity reflecting system's selective stress reaction toward respective threat" (Vamanu et al., 2016, p. 95). Details aside, that model's "assumption zero" is that critical, or otherwise, complex real-life structures and systems can be accommodated within the concept of a multi-component, multi-indicator system, the parts of which would show some kind of collective behavior by their interactions as well as some susceptibility to external factors acting upon the structure as a whole. To quantify the vulnerability of such a generic system, the model rests on two control variables and an equation of state. The model input includes an arbitrarily large number of indicators accounting for the system's internal processes (fast-varying) and external forces (slow-varying) assumed to act upon the components of the entire structure uniformly. The model output is a membership fraction qualifying the integrity of the system in terms of operability defined in terms of the proportion of "operable" versus "inoperable" states of the system. The solution is directly conducive to the definition of a *two-parameter phase space* of the system, where three *vulnerability basins* may be identified: (i) system stable—low vulnerability; (ii) system and vulnerability—critical; and (iii) system unstable—high vulnerability. Moreover, a 0–100 *vulnerability scale* and the means to measure the respective *vulnerability index* was offered as an operational expression of a QVA.

Talking about "basins of vulnerability" highlights an important issue: tolerating vulnerability. After all, people may tolerate low (or acceptable) vulnerability and reject critical (or unacceptable) high vulnerability—an issue pointing at stakeholder perception of vulnerability. If vulnerability can be quantified, it can be graded, mapped, and managed, and policies of making systems can even be conceived, including supply chain and logistics networks, only *as resilient as reasonably achievable* (ARARA). Therefore, developing a resilient supply chain and logistics network requires consideration of vulnerability in some form or another.

1.2.3 System Perception

Perception plays an important role in everyday life, including the view of risk and vulnerability. Concerning a risk, Hill (2012) notes that an analyst's perception of the probability of occurrence of an event and the bearing of its potential consequences are fundamental for one's understanding and value, threat, vulnerability, and consequence. Notice that perception is intrinsic to an individual and is therefore related to deep-seated fundamental assumptions such as one's beliefs and predispositions. In examining a well-known disaster (i.e., Hurricane Katrina), a sobering finding is inescapable: *it is possible to make decisions and take actions contrary to what is expected.* Why one *might not* evacuate, despite the high probability of occurrence of an event and heavy consequences, the answer remains hidden in the perception of the individuals (Katina, 2016). In this regard, Hill (2012) suggests that it is possible to have "subjective judgment about the severity of a risk scenario to an asset; [it] may be driven by sense, emotion, or personal experience" (Hill, 2012, p. 20).

This idea is also supported in Reason's (1990) research, suggesting that making decisions and taking actions are related to mental cognitive processes. Moreover, one's background affects perception, as in the case involving a survey that seems to indicate that, contrary to their American counterparts, Chinese students tended to engage in "risky" behavior since their culture emphasizes collectivism and interdependence in family and the community as a whole (Weber & Hsee, 1998). Advocating for the inclusion of perception are models for analyzing social (e.g., Social Amplification of Risk Framework) and culture (i.e., Cultural Theory Model). For example, the Cultural Theory Model assumes that people choose to worry about certain risk scenarios based on their social engagements (Sjöberg, 1999). Therefore, while developing a resilient supply chain and logistics network requires the consideration of risk and vulnerability in some form or another, it also appears there is a need to consider perception.

1.2.4 SYSTEM FRAGILITY

In describing the fragility of a system, from the perspective of complex adaptive systems, Johnson and Gheorghe posit the following about a dynamic environment (Johnson & Gheorghe, 2013, pp. 160–161): "a host of things are always changing: conditions, constraints, treats, opportunities, and so on. The ability to make internal adjustments in response to, or in anticipation of, external environmental changes, is the essence of being adaptive. In less complex systems, these changes take place based on pre-established rules in the system . . . [however]. Complex adaptive systems . . . are not only responsive to environmental dynamics; they have the ability to learn from experiences." Subsequently, how complex systems respond to hazards can be characterized on a continuum ranging from "fragile" to "robust" to "antifragile" (Johnson & Gheorghe, 2013). In this case, fragility indicates the system's possibility of being degraded by stress/threat. "Robust" means that the system remains unchanged when under stress/threat while "antifragility" means the ability to improve with stress (Taleb, 2014). On the surface, one could submit that vulnerability and fragility appear to describe a similar consideration of a system. However, there is a critical distinction for those interested in examining system *failures*, relating to why a system can become vulnerable or fragile. Johnson and Gheorghe (2013) offer a response: "Vulnerable systems fail because of their degree of exposure to a stress [hazard] of a specific nature, while fragile systems fail because they are easily broken regardless of the nature of stress they are exposed to" (Johnson & Gheorghe, 2013, p. 161).

More interesting is the "anti" fragility. Taleb (2014) argues that to a certain degree, the ability of a system to withstand stress is a function of some deliberate, intentional exposure to small stressing events. In effect, one can achieve antifragility by introducing small stresses, thereby strengthening the system against extreme stress. While this idea might sound counterintuitive at first glance, exposure to stress, especially at an early stage, is beneficial in preparing systems for future stressful events (Becvar & Becvar, 1999; Herzfeld, 2001; Katina, 2015).

When the potential benefits of antifragility are considered, it becomes evident that supply chain and logistics networks might need to be exposed intentionally to low-level stressing events to increase their ability to withstand higher-level or even extreme stresses. Such endeavors would involve sophisticated methods and tools to manage fragility and increase antifragility levels properly. While such approaches might take advantage of advanced simulation techniques (e.g., quantum cellular automata and parallel computing) to simulate stress and its effects on supply chains, the case is made that stress must involve risk, vulnerability, and perception.

1.2.5 CONCATENATION FOR SYSTEM RESILIENCE

Vulnerability analysis of a system can rest on two interrogations: First, how *susceptible* is the system to suffer from threats if threats materialize? And second, to what degree can a system *recover* from an effective threat hit and how fast? The second implies *how resilient the system is*. If one takes susceptibility as the amount of damage per unit of hit intensity expressed in, for example, monetary terms, and resilience may be seen as a Cartesian product of an *acceptable* percent recovery of system operability by the time lapse to reach that percentage, then both susceptibility and resilience can be convincingly quantified.

Vulnerability could be taken as a function of susceptibility and resilience (Gheorghe et al., 2018; Song, 2005). However, this view may deprive vulnerability of many complexities that make it worth recognition and appreciation. Nevertheless, leaving aside a scholastic debate and focusing on the "accepted" perception that resilience is the *ability to withstand a threat/attack* (Gheorghe & Katina, 2014; Martin-Breen & Anderies, 2011), one also has to recognize that withstanding an attack can unambiguously be equivalent to *recover from the hit up to an acceptable, fractional level from its original, nominal operability*. The concept of resilience is important in the practice of supply chain management for several reasons, among which one may stand out: *a system may be susceptible and yet have a high resilience*. And from a vulnerability standpoint, the latter feature is overwhelmingly consequential.

The supply chain and logistics network, especially from the point of view of goods and services (including reliefs), calls for recovery as soon as possible. Failure to return to some degree of normality can easily lead to crisis and a debilitating impact. This was the case during the California electricity crisis of 2000–2001 in which the state suffered large-scale blackouts, followed by political turmoil and electricity supply shortages rooted in market manipulations, the illegal shutdown of pipelines, and capped retail electricity prices (Sweeney, 2002).

Moreover, in dealing with supply chain and logistics networks, susceptibility does not equate to performance reduction. A network could be a susceptible system and remain highly resilient and, thereby, classified as an acceptable level of vulnerability—undoubtedly implying that specific mechanisms to enhance system capabilities to withstand hazards have implications on solutions to provide the necessary resources to recover promptly.

Indeed, in a broader sense, interest in supply chain and logistics network resilience can be concerned with evaluating existing resilience mechanisms—technical, logistic, and operational—relating to selected potential disasters. Such considerations should provide the basis for developing prioritization and benefit trade-offs for investments to increase resilience, preparedness, and response capabilities for a given system of interest and even regions. Moreover, prioritization and ranking can also serve as the basis for identifying the most critical scenarios related to the unwanted situation, ranking the vulnerable points to provide a foundation for allocation of limited resources (and establishment of mitigation plans) and determining the dominating (and evolving) vulnerabilities for further assessment (Song, 2005).

In such an analysis, *sensitivity* and *attractiveness* can serve as surrogate measures for the likelihood of disruptive events (Vamanu et al., 2016). For sensitivity, there is a need to examine whether the vulnerable points are under critical stress under various internal (or external) unfavorable conditions. For attractiveness, one assumes that the more vulnerable a point, the more it makes an "attractive" target for the adversary. In this case, an adversary can be of a natural (random, unintentional) or anthropic (intentional, malevolent) origin. In this discussion, it should be apparent that those in charge of designing our supply chains must strive to improve the defensive properties of such systems. An in-depth investigation of such matters must yield aspects of network system protective "ilities", including the likes of adaptability, availability (e.g., of warning systems), detection, deterrence, maintenance, redundancy, reliability, and robustness.

In summary, the practice of supply chain and logistics network, while desirable and capable as is, can benefit from the ideas of resilience, especially in the complexity and ambiguous landscape. Risk, vulnerability, perceptions, and fragility run wild in such situations. The following section explores the possible role of technology in creating a resilient supply chain and logistics network.

1.3 BLOCKCHAIN TECHNOLOGY

The use of technology to tackle challenges is not new. However, the use of blockchain technology coupled with the concept of resilience to address challenges in disasters is novel. Blockchain is a technology. This technology can be described as a tool comprised of a distributed ledger that "keeps track of all transactions that have taken place across a peer-to-peer network" (Keerati, 2017, p. 4). Google's Crosby and his colleagues at Yahoo, Samsung, and Fairchild Semiconductor are convinced that once transactions are entered into this ledger and verified by the consensus of a majority of the participants in the system, they can never be erased (Crosby et al., 2016).

Blockchain is formally defined as a "fully distributed system for cryptographically capturing and storing consistent, immutable, linear event log for transactions between network participants" (Risius & Spohrer, 2017, p. 386). Transparency is maintained within the system through consensus (Queiroz et al., 2019). The cryptographically linked blocks of transactions form a blockchain (Kshetri, 2017) that occurs via hash functions (Casado-Vara et al., 2018). Hash is a unique code to identify each block (Moosavi et al., 2021), and hashing is critical for the immutability of the blockchain.

1.3.1 FOUNDATIONS OF BLOCKCHAIN TECHNOLOGY

Blockchain is probably best known as a technology that underpins Bitcoin cryptocurrency, taking records (e.g., proofs of ownerships, confirmed financial transactions, and financial contracts) and placing them into "blocks," which are linked to prior blocks—hence forming a "chain" of blocks chronologically (Alvseike & Iversen, 2017; CBS Interactive, 2018; Crosby et al., 2016; Kakushadze & Russo, 2018; Keerati, 2017).

Nodes in the blockchain communicate via the network by standard communication protocols. Every block with the chain network needs a dispersed consensus mechanism. The consensus function provides immutability by verifying the network transactions (Wang et al., 2018). New transactions are added to the blockchain by miners using a consensus algorithm that should be confirmed by most nodes of the network via a voting operation. As the network approves the transactions, they permanently become valid and part of the database. The system rewards miners for adding valid blocks to constantly validate and maintain consistent data by spending their computing power (Tian, 2017). A valid block is generated using a consensus algorithm, which is a challenging puzzle to be solved and requires a massive amount of computational cost but is easy to verify. The mining node broadcasts the completed blocks. The most notable consensus algorithms are proof of work, proof of stake, practical Byzantine fault tolerance, Ripple, and Tendermint (Gausdal et al., 2018).

The blocks are collected in a chain, verified, and managed via governance protocols. Every party can verify the records without a distributed consensus mechanism or an intermediary. The modern encryption methods and verification process secure the data on the ledgers against manipulation. Users have access to the audit trail of actions. The decentralized data storage decreases any single point's failure risk (Wang et al., 2018). Figure 1.3 summarizes blockchain operations (transactions, verification, and blocks), adapted from Yoo and Won (2018).

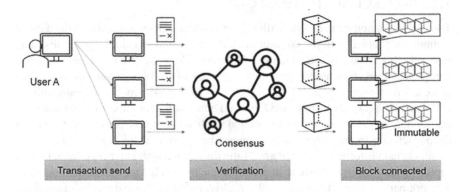

FIGURE 1.3   The three types of operations for blockchains.

Literature implies that decentralization, disintermediation, immutability, creation (movement of digital assets), and transaction sharing as key aspects of blockchain technology (Queiroz et al., 2019; Tian, 2017). With the blockchain implementation, trustless networks are developed, which do not need trust for transferring to other users. The disintermediation feature makes transactions faster between users. Also, information is secure in the blockchain with cryptography. Ivanov et al. (2019)'s research further suggests that data integrity and the distributed nature of blockchain enable participants to transact with a high level of confidence. Making any changes in a backdated transaction would be very complicated since it requires modifying all of the following blocks (Guan et al., 2018). The modulation of the block needs more time than the verification, which makes forging and falsifying the data difficult (Yoo & Won, 2018). The scalability of blockchain technology can be realized by five planes of network, consensus, storage, view, and side planes, which are dependent in order from bottom to top (Tian, 2017).

A related feature of blockchain is a smart contract. A smart contract is a computer software program that stores procedures and regulations for negotiating terms and actions between participants (Casado-Vara et al., 2018). Nick Szabo introduced the smart contract concept in 1994 as a "computerized transaction protocol that executes the terms of a contract" (Casado-Vara et al., 2018). Yoo and Won (2018) define a smart contract as "a protocol that automates and replaces the necessity of a contract, such as negotiation, expediting performance, confirmation, and contract clauses on execution." Wang et al. (2018) suggest that a smart contract is a computerized transaction protocol that automatically executes the term of a contract upon a blockchain. Smart contract applications can reduce expenses as well as delays when compared to traditional contracts, and as such expedites common contractual circumstances. Smart contracts can represent business logic, mechanisms, or decision-making and communicate with participants. Hence, smart contracts create a web of tiny services that enable the creation and autonomous activity of complex systems (Glaser, 2017). Smart contracts offer significant support to the critical activities across industry sectors, including supply chain, finance, and medical services (Tolmach et al., 2021). The network participants can reach a consensus on the outcome of the contract execution. Transactions in the smart contract are implemented autonomously and automatically in a prescribed way on every node of the network based on pre-established protocols (Casado-Vara et al., 2018).

1.3.2 INSTANTIATIONS OF BLOCKCHAIN TECHNOLOGY

Again, blockchain is probably best known as a technology that underpins Bitcoin cryptocurrency. Created by Satoshi Nakamoto in 2008, Bitcoin is a type of unregulated digital currency (Nakamoto, 2008). It is a form of cryptocurrency, a digital asset designed to work as a medium of exchange that uses strong cryptography to secure financial transactions, control the creation of additional units, and verify the transfer of assets. It was launched to bypass government currency control mechanisms and simplify online transactions by eliminating third-party payment processing intermediaries. In contrast, blockchain is a technology that works as a distributed ledger that can keep track of all transactions on a network. Bitcoin Blockchain is only one instantiation of blockchain technology. Blockchain technology can be implemented in several ways, including "traditional" (and side), mutable data (and immutable data), public (and private), and transparent data (and opaque data) blockchains.

1.3.2.1 Traditional and Side

A traditional blockchain can be defined as a ledger keeping track of all transactions across a loosely "closed" network. If two different organizations, each operating on an internal network, want to communicate, then they can evoke a "side" chain (Katina et al., 2019). A sidechain is a type of blockchain that allows for data transfer among traditional blockchains "without breaking the immutability property imposed by the underlying technology" (Keerati, 2017, p. 11). Bitcoin represents a use case for sidechains. Organizations can create internal Bitcoin networks carved out from the public network, allowing for internal privacy (Keerati, 2017).

1.3.2.2 Immutable and Mutable Data

A blockchain can also be designed in a manner it records transactions: permanent or modifiable. A permanent blockchain is immutable—once a transaction has been committed, it cannot be reversed or changed (Keerati, 2017). However, one can append the chain by creating a new block. Bitcoin is an example of an immutable blockchain where new transactions can be appended to the Bitcoin blockchain. Mutable blockchains allow for modifications of existing data. These modifications still have to go through the consensus approval process. Accenture Blockchain and IBM Blockchain are representatives of mutable blockchains for business in which modifications of existing data are possible, offering more efficiency as opposed to security in the immutable blockchains.

1.3.2.3 Public and Private

A public blockchain is a permission-less blockchain that can grant read and write access to all users who wish to join the network. A public blockchain allows for broader access. However, they are harder to control for privacy and harder to apply in a specialized mode. Examples include Bitcoin, Ethereum, Dash, and Lisk. A private blockchain allows only permitted parties to join the network. These blockchains could be designed for specific purposes. However, they can limit access, making permitted parties intermediaries for other networks outside the network. Examples include Ripple, R3 Corda, and Chain.

1.3.2.4 Transparent and Opaque Data

A blockchain can also be described in terms of transparency and privacy. When "data on blockchain is encrypted, blockchain can still be designed for different levels of transparency and degrees of privacy. Some blockchains—such as Bitcoin—are designed so that one can still identify the parties engaging in transactions via pseudonyms. In certain instances, one can then use network analysis to decipher the pseudonyms and reveal the actual identities of the parties engaging in transactions" (Keerati, 2017, p. 8). An *opaque blockchain* is "darkened" for all parties to engage in transactions under total privacy. An exemplary example includes Z-cash, which allows parties to engage in transactions under "zero-knowledge security."

However, optional "selective disclosure" allows a user to prove payment for auditing purposes.

1.3.3 BENEFITS OF BLOCKCHAIN TECHNOLOGY

Beyond the excitement of using technology to tackle a problem, there are varying perspectives on blockchain technology's benefits (and threats)—we will try and skip threats. Still, we suggest Abu-elezz et al. (2020)'s research for those interested.

1.3.3.1 Elimination of Single Point of Failure

A single failure can halt operations as a whole interconnected system. This potential can be reduced via the network approach built on blockchain technology via the role of intermediaries. Potential risks of failure (e.g., through mass demand, security attack, or other technical glitches) at intermediary nodes are eliminated via the blockchain structure. In a blockchain, even if a node fails, other nodes can still maintain the records and verify all the transactions on the network (Keerati, 2017).

1.3.3.2 Reduction of Record-Keeping Costs

A blockchain can reduce record-keeping costs by eliminating the need for the reconciliation process and the risks of a "double spending." Additionally, participants in a blockchain network share the same distributed ledger. Since transactions can occur in real time, there is no need to reconcile records among counterparties. These parties also don't need to worry that their counterparties will engage in multiple transactions using the same assets. Finally, the auditing processes are enhanced via smart contracts.

1.3.3.3 Reduction of Risk of Fraud and Information Leakage

A blockchain database can be encrypted and immutable. In this case, accessing and manipulating data requires a "public key" and a "private key," which help ensure data security. More importantly, any changes to the data require verification by participants in the network. Thus, it is virtually impossible to manipulate the data on a decentralized blockchain system, even for seasoned hackers (Katina & Keating, 2018; Keerati, 2017).

1.3.3.4 Removal of Transaction Intermediaries

Research suggests that one of the unique benefits of blockchain is facilitating secure, decentralized transactions among unrelated parties without going via intermediaries (Keerati, 2017). In many cases, intermediaries increase the complexity of the transactions. The structure of a blockchain facilitates the removal of intermediaries between parties. This means, for example, an exchange of funds without a bank/broker is now possible.

1.3.3.5 Transaction Time Reduction

A blockchain has the capability to reduce the time it takes to accomplish certain activities. There is no need for transaction reconciliation as transactions are conducted on a shared ledger. This shared ledger is visible to all parties. This means, for example, it is possible to do settlements in nanoseconds, just as it is done in the trading of stocks and bonds.

In the interest of the present motivations, research also suggests that blockchain has the potential to support the achievement of critical supply chain objectives (Kshetri, 2018). For example, eliminating intermediaries through smart contract means automated asset transfer if the determining conditions are fulfilled based on the blockchain decentralization feature. These features of blockchain can be used to support supply chain and logistics network management innovation and reconfiguration (Queiroz & Fosso Wamba, 2019). Blockchain can be applied to register time, location, price, involved parties, and related information while the ownership of an item is changing (Kshetri, 2017). Saberi et al. (2019) also suggest that challenges (e.g., inefficient transactions, fraud, pilferage, and poor performance) can be eliminated by zero trust, accurate information sharing, and verifiability in blockchains.

Moreover, the technological developments and applications of blockchain technology can improve supply chain transparency, security, durability, and process integrity in organizational, technological, and economic feasibility (Saberi et al., 2019). As the supply networks contain large numbers of stakeholders, tracking processes becomes more difficult. Smart contracts can automate the processes. The agreed contracts can be delivered to the specified parties for digital execution, programs can be updated based on agreed verifications, and copyright documents can be released to the relevant parties. The adoption of smart contracts can fundamentally change the supply chain structures and governance (Wang et al., 2018). Smart contracts can facilitate supply chain monitoring and control. The logistics planning, commercial contracts, and requirements of customers automatically and efficiently can be transmitted from retailers to manufacturers and suppliers (Chen et al., 2017). Smart contracts impact data sharing among supply network participants and provide continuous process improvement (Saberi et al., 2019).

1.3.4 TOWARDS BLOCKCHAIN-ENABLED RESILIENCE

A disaster can be defined as a severe problem occurring over a short or long period of time that causes widespread human, material, and economic or environmental loss that exceeds the ability of the affected community or society to cope using its resources (IFRC, 2022; WHO/EHA, 2002). Suffice it to say that at the heart of any disaster is the need for recovery and relief—this forms an essential aspect of organized supply chain management activities.

Substantial operations, unique constraints, irregular demand, and erratic supply and transportation information define disaster supply chain and logistics management. In essence, the disaster area is a dynamic environment. The blockage can be in supply distribution coming from different sources, including damaged infrastructure affecting the availability and accuracy of information. Success in the disaster supply chain and logistics network is primarily based on the collaboration and coordination of involved parties, especially when distributing scarce resources. Managers in such situations are expected to make good decisions (transparent, trusted, and traceable), and as such, they must be able to balance the uncertainty in the decision-making process.

Moreover, there is a need to evaluate, integrate, and harmonize the activities of the various parties that emerge to deal with the incidents (Day et al., 2012). Balcik and Beamon (2008) also support this view, suggesting that there can also be a need to face the unknowns such as culture and politics. In dealing with disasters, especially the flow of resources in a disaster area, Holguín-Veras et al. (2007) suggest a sequential process involving assessment, deployment, sustainment, and reconfiguration. The reality is that a multidisciplinary investigative approach is obligatory to effectively address emergencies involving multiple agencies with many uncertainties (AliPour, 2021). Additionally, disasters tend to be dynamic and involve time-sensitive operations, particularly in the initial phase. Research also indicates that equity and fairness among aid beneficiaries, in these situations, require more consideration so as not to deprive the needed relief to the destitute (Cavdur et al., 2016; Tierney, 2012).

A promising technological innovation of blockchain technology promises real-time, secure, and transparent information exchange as well as automation through smart contracts. Moreover, blockchain technology and the Internet of Things (IoT) can be combined to enable dynamic, reliable, and transparent tracking of supply chains (Betti et al., 2020; Pérez-Solà et al., 2019).

As a field, disaster management is vast and involves the design and management of current as well as future disasters. Moreover, disaster management is also responsible for identifying, implementing, and monitoring the desired outcomes. The present study focuses on the information and communication influences on disaster management within (i) the increasing volatility, uncertainty, complexity, and ambiguity (characteristics that tend to define recent disasters) and (ii) the dawn of blockchain technologies and simulation (which renders an opportunity to redefine how to deal with the disaster ecosystem).

The study's focus is a good challenge unless, of course, one still believes disasters are calamities to be blamed on the position of planets, just as in fact the old world believed (Harper, 2022).

CONCLUSIONS

This chapter articulates the fundamentals of supply chain management along with the need for effective methods and tools to address increasing complexity and ambiguity in the operational landscape of supply chain and logistics networks. Supply chain management must be systemic and coordinate the traditional business functions strategically. It includes tactics for across-business functions within a particular company and across businesses within the supply chain to improve short and long-term performance. However, the supply chain management practice must also account for emerging issues such as the need for resiliency and transparency, especially in crises and disasters.

The use of new technologies to tackle challenges is not new. However, the use of blockchain technology and the concept of resilience to address disaster challenges is novel. Blockchain technology offers the ability to inform the ledger with consent, rapidly share the transactions, and record time-stamped and tamper-proof auditable accounts. Since resilience involves the ability to withstand, recover from, and reorganize in response to crises, infusing blockchain and ideas of resilience can enhance supply chains and logistics in disaster relief by affecting disaster recovery plans, including transparency, efficiency, scale, and sustainability.

A starting point might be the development of the blockchain-enabled resilience model. The development, application, implications, and research agenda are the basis for the remainder of this book.

REFERENCES

Abu-elezz, I., Hassan, A., Nazeemudeen, A., Househ, M., & Abd-alrazaq, A. (2020). The benefits and threats of blockchain technology in healthcare: A scoping review. *International Journal of Medical Informatics*, *142*, 104246. https://doi.org/10.1016/j.ijmedinf.2020.104246

Alberts, C. J., Haller, J., Wallen, C. M., & Woody, C. (2017). Assessing DoD system acquisition supply chain risk management. *CrossTalk*, *30*(3), 4–8.

Alvseike, R., & Iversen, G. A. G. (2017). *Blockchain and the future of money and finance: A qualitative exploratory study of blockchain technology and implications for the monetary and financial system* [Masters, Norwegian School of Economics]. https://brage.bibsys.no/xmlui/handle/11250/2453330

ASCE. (2009). *Guiding principles for the nation's critical infrastructure*. American Society of Civil Engineers. www.asce.org/uploadedFiles/Infrastructure_-_New/GuidingPrinciples-FinalReport.pdf

ASCM. (2022). *APICS dictionary: The essential supply chain reference*. Association for Supply Chain Management. www.ascm.org/learning-development/certifications-credentials/dictionary/

Atkins, E. (2020, April 3). Keeping up: How to keep your DC moving during the pandemic. *Inside Logistics*. www.insidelogistics.ca/features/keeping-up-how-to-keep-your-dc-moving-during-the-pandemic/

Aven, T. (2011). On some recent definitions and analysis frameworks for risk, vulnerability, and resilience. *Risk Analysis*, *31*(4), 515–522. https://doi.org/10.1111/j.1539-6924.2010.01528.x

Balcik, B., & Beamon, B. M. (2008). Facility location in humanitarian relief. *International Journal of Logistics Research and Applications*, *11*(2), 101–121. https://doi.org/10.1080/13675560701561789

Bales, R. R., Maull, R. S., & Radnor, Z. (2004). The development of supply chain management within the aerospace manufacturing sector. *Supply Chain Management: An International Journal*, *9*(3), 250–255. https://doi.org/10.1108/13598540410544944

Basole, R. C., Bellamy, M. A., Park, H., & Putrevu, J. (2016). Computational analysis and visualization of global supply network risks. *IEEE Transactions on Industrial Informatics*, *12*(3), 1206–1213. https://doi.org/10.1109/TII.2016.2549268

Beamon, B. M. (1999). Measuring supply chain performance. *International Journal of Operations & Production Management*, *19*(3), 275–292. https://doi.org/10.1108/01443579910249714

Becvar, D. S., & Becvar, R. J. (1999). *Systems theory and family therapy: A primer* (2nd ed.). University Press of America.

Betti, Q., Montreuil, B., Khoury, R., & Hallé, S. (2020). Smart contracts-enabled simulation for hyperconnected logistics. In M. A. Khan, M. T. Quasim, F. Algarni, & A. Alharthi (Eds.), *Decentralised internet of things: A blockchain perspective* (pp. 109–149). Springer International Publishing. https://doi.org/10.1007/978-3-030-38677-1_6

Blanchard, B. S. (2008). *System engineering management*. John Wiley and Sons, Inc.

Casado-Vara, R., Prieto, J., la Prieta, F. D., & Corchado, J. M. (2018). How blockchain improves the supply chain: Case study alimentary supply chain. *Procedia Computer Science, 134*, 393–398. https://doi.org/10.1016/j.procs.2018.07.193

Cavdur, F., Kose-Kucuk, M., & Sebatli, A. (2016). Allocation of temporary disaster response facilities under demand uncertainty: An earthquake case study. *International Journal of Disaster Risk Reduction, 19*, 159–166. https://doi.org/10.1016/j.ijdrr.2016.08.009

CBS Interactive. (2018). *Blockchain: An insider's guide*. Tech Republic. www.techrepublic. com/resource-library/whitepapers/blockchain-an-insider-s-guide/

Chen, S., Shi, R., Ren, Z., Yan, J., Shi, Y., & Zhang, J. (2017). A blockchain-based supply chain quality management framework. *2017 IEEE 14th International Conference on E-Business Engineering (ICEBE)*, 172–176. https://doi.org/10.1109/ICEBE.2017.34

Cornell Engineering. (2022). *Supply chain*. Operations Research and Information Engineering. www.orie.cornell.edu/supply-chain

Craighead, C. W., Blackhurst, J., Rungtusanatham, M. J., & Handfield, R. B. (2007). The severity of supply chain disruptions: Design characteristics and mitigation capabilities. *Decision Sciences, 38*(1), 131–156. https://doi.org/10.1111/j.1540-5915.2007.00151.x

Crosby, M., Nachiappan, Pattanayak, P., Verma, S., & Kalyanaraman, V. (2016). BlockChain technology: Beyond bitcoin. *Applied Innovation Review, 2*, 6–19.

Day, J. M., Melnyk, S. A., Larson, P. D., Davis, E. W., & Whybark, D. C. (2012). Humanitarian and disaster relief supply chains: A matter of life and death. *Journal of Supply Chain Management, 48*(2), 21–36. https://doi.org/10.1111/j.1745-493X.2012.03267.x

Einarsson, S., & Rausand, M. (1998). An approach to vulnerability analysis of complex industrial systems. *Risk Analysis, 18*(5), 535–546. https://doi.org/10.1111/j.1539-6924.1998. tb00367.x

Fawcett, S. E., & Magnan, G. M. (2002). The rhetoric and reality of supply chain integration. *International Journal of Physical Distribution & Logistics Management, 32*(5), 339–361. https://doi.org/10.1108/09600030210436222

García-Herrero, A. (2020, February 17). Epidemic tests China's supply chain dominance. *European*. https://european.economicblogs.org/bruegel/2020/garc%c3%ada-herrero-epidemic-tests-china-supply-chain-dominance

Garvey, P. R. (2009). *Analytical methods for risk management: A systems engineering perspective*. Chapman & Hall/CRC.

Gausdal, A. H., Czachorowski, K. V., & Solesvik, M. Z. (2018). Applying blockchain technology: Evidence from Norwegian companies. *Sustainability, 10*(6), 1985. https://doi. org/10.3390/su10061985

Gheorghe, A., Vamanu, D. V., Katina, P., & Pulfer, R. (2018). *Critical infrastructures, key resources, key Assets: Risk, vulnerability, resilience, fragility, and perception governance* (Vol. 34). Springer International Publishing. www.springer.com/us/book/9783319692234

Gheorghe, A. V., & Katina, P. F. (2014). Editorial: Resiliency and engineering systems— Research trends and challenges. *International Journal of Critical Infrastructures, 10*(3/4), 193–199.

Gibson, J. E., Scherer, W. T., & Gibson, W. F. (2007). *How to do systems analysis*. Wiley-Interscience.

Glaser, F. (2017, January 1). Pervasive decentralisation of digital infrastructures: A framework for blockchain enabled system and use case analysis. *Proceedings of the 50th Hawaii International Conference on System Sciences (HICSS-50)*. https://papers.ssrn.com/ abstract=3052165

Golicic, S. L., & Davis, D. F. (2012). Implementing mixed methods research in supply chain management. *International Journal of Physical Distribution & Logistics Management, 42*(8/9), 726–741. https://doi.org/10.1108/09600031211269721

Griffy-Brown, C. (2003). Just-in-time to just-in-case. *Graziadio Business Review, 6*(2), 1–4.

Guan, Z., Si, G., Zhang, X., Wu, L., Guizani, N., Du, X., & Ma, Y. (2018). Privacy-preserving and efficient aggregation based on blockchain for power grid communications in smart communities. *IEEE Communications Magazine*, *56*(7), 82–88. https://doi.org/10.1109/MCOM.2018.1700401

Harper, D. (2022). Disaster. In *Online etymology dictionary*. Douglas Harper. www.etymonline.com/word/disaster

Haskins, C. (2006). 8.1.1 Application of systems engineering to industrial supply chains. *INCOSE International Symposium*, *16*(1), 1093–1109. https://doi.org/10.1002/j.2334-5837.2006.tb02798.x

Herzfeld, M. (2001). *Anthropology: Theoretical practice in culture and society*. Blackwell Publishers.

Hill, K. N. (2012). *Risk quadruplet: Integrating assessments of threat, vulnerability, consequence, and perception for homeland security and homeland defense* [PhD, Old Dominion University]. http://search.proquest.com.proxy.lib.odu.edu/pqdtlocal1005724/docview/1034365511/abstract/3E84BAAB80A043B7PQ/14

Holguín-Veras, J., Pérez, N., Ukkusuri, S., Wachtendorf, T., & Brown, B. (2007). Emergency logistics issues affecting the response to Katrina: A synthesis and preliminary suggestions for improvement. *Transportation Research Record*, *2022*(1), 76–82. https://doi.org/10.3141/2022-09

Holmgren, A., Molin, S., & Thedéen, T. (2001). Vulnerability of complex infrastructure; power system and supporting digital communication system. 5th International Conference on Technology, Policy, and Innovation, Utrecht, the Netherlands.

Holton, G. A. (2004). Defining risk. *Financial Analysts Journal*, *60*(6), 19–25.

Huan, S. H., Sheoran, S. K., & Wang, G. (2004). A review and analysis of supply chain operations reference (SCOR) model. *Supply Chain Management: An International Journal*, *9*(1), 23–29. https://doi.org/10.1108/13598540410517557

IFRC. (2022, May 26). *What is a disaster?* International Federation of Red Cross and Red Crescent Societies. www.ifrc.org/what-disaster

INCOSE. (2011). *Systems engineering handbook: A guide for system life cycle processes and activities* (H. Cecilia, Ed.; 3.2). INCOSE.

IRGC. (2007). *Managing and reducing social vulnerabilities from coupled critical infrastructures* (pp. 1–16). International Risk Governance Council. www.irgc.org/IMG/pdf/IRGCinfra_site06.11.07-2.pdf

Ivanov, D., Dolgui, A., & Sokolov, B. (2019). The impact of digital technology and Industry 4.0 on the ripple effect and supply chain risk analytics. *International Journal of Production Research*, *57*(3), 829–846. https://doi.org/10.1080/00207543.2018.1488086

Johnson, J., & Gheorghe, A. V. (2013). Antifragility analysis and measurement framework for systems of systems. *International Journal of Disaster Risk Science*, *4*(4), 159–168. https://doi.org/10.1007/s13753-013-0017-7

Kakushadze, Z., & Russo Jr, R. P. (2018). Blockchain: Data malls, coin economies and keyless payments. *The Journal of Alternative Investments*, *21*(1), 8–16. https://arxiv.org/ftp/arxiv/papers/1802/1802.07422.pdf

Kamble, S. S., & Gunasekaran, A. (2020). Big data-driven supply chain performance measurement system: A review and framework for implementation. *International Journal of Production Research*, *58*(1), 65–86. https://doi.org/10.1080/00207543.2019.1630770

Katina, P. F. (2015). *Systems theory-based construct for identifying metasystem pathologies for complex system governance* [PhD, Old Dominion University].

Katina, P. F. (2016). Individual and societal risk (RiskIS): Beyond probability and consequence during Hurricane Katrina. In A. J. Masys (Ed.), *Disaster forensics: Understanding root cause and complex causality* (pp. 1–23). Springer International Publishing. https://doi.org/10.1007/978-3-319-41849-0_1

Katina, P. F., & Keating, C. B. (2018). Cyber-physical systems governance: A framework for (meta)cybersecurity design. In A. J. Masys (Ed.), *Security by design: Innovative perspectives on complex problems* (pp. 137–169). Springer International Publishing. https://doi.org/10.1007/978-3-319-78021-4_7

Katina, P. F., Keating, C. B., Sisti, J. A., & Gheorghe, A. V. (2019). Blockchain governance. *International Journal of Critical Infrastructures*, *15*(2), 121. https://doi.org/10.1504/IJCIS.2019.098835

Katina, P. F., Pinto, C. A., Bradley, J. M., & Hester, P. T. (2014). Interdependency-induced risk with applications to healthcare. *International Journal of Critical Infrastructure Protection*, *7*(1), 12–26. https://doi.org/10.1016/j.ijcip.2014.01.005

Keating, C. B., & Katina, P. F. (2019). Complex system governance: Concept, utility, and challenges. *Systems Research and Behavioral Science*, *36*(5), 687–705. https://doi.org/10.1002/sres.2621

Keerati, R. (2017). *Preparing for blockchain: Challeges and alternatives for financial regulators*. University of California. https://ctsp.berkeley.edu/preparing-for-blockchain/

Knight, F. H. (1921). *Risk, uncertainty, and profit*. Hart, Schaffner & Marx; Houghton Mifflin Co. http://hdl.loc.gov/loc.gdc/scd0001.00019647742

Kozlenkova, I. V., Hult, G. T. M., Lund, D. J., Mena, J. A., & Kekec, P. (2015). The role of marketing channels in supply chain management. *Journal of Retailing*, *91*(4), 586–609. https://doi.org/10.1016/j.jretai.2015.03.003

Kshetri, N. (2017). Blockchain's roles in strengthening cybersecurity and protecting privacy. *Telecommunications Policy*, *41*(10), 1027–1038. https://doi.org/10.1016/j.telpol.2017.09.003

Kshetri, N. (2018). 1 Blockchain's roles in meeting key supply chain management objectives. *International Journal of Information Management*, *39*, 80–89. https://doi.org/10.1016/j.ijinfomgt.2017.12.005

Lam, H. K. S. (2018). Doing good across organizational boundaries: Sustainable supply chain practices and firms' financial risk. *International Journal of Operations & Production Management*, *38*(12), 2389–2412. https://doi.org/10.1108/IJOPM-02-2018-0056

Lambert, D. M., & Cooper, M. C. (2000). Issues in supply chain management. *Industrial Marketing Management*, *29*(1), 65–83. https://doi.org/10.1016/S0019-8501(99)00113-3

Ludema, Ir. M. W. (2002). 2.6.5 Designing a supply chain analysis framework. *INCOSE International Symposium*, *12*(1), 1092–1099. https://doi.org/10.1002/j.2334-5837.2002.tb02577.x

Martin-Breen, P., & Anderies, J. M. (2011). *Resilience: A literature review* (p. 64). The Rockefeller Foundation. www.rockefellerfoundation.org/blog/resilience-literature-review

Mentzer, J. T., DeWitt, W., Keebler, J. S., Min, S., Nix, N. W., Smith, C. D., & Zacharia, Z. G. (2001). Defining supply chain management. *Journal of Business Logistics*, *22*(2), 1–25. https://doi.org/10.1002/j.2158-1592.2001.tb00001.x

Moosavi, J., Naeni, L. M., Fathollahi-Fard, A. M., & Fiore, U. (2021). Blockchain in supply chain management: A review, bibliometric, and network analysis. *Environmental Science and Pollution Research International*. https://doi.org/10.1007/s11356-021-13094-3

Nakamoto, S. (2008). *Bitcoin: A peer-to-Peer electronic cash system*. https://bitcoin.org; https://bitcoin.org/bitcoin.pdf

Neely, A., Gregory, M., & Platts, K. (1995). Performance measurement system design: A literature review and research agenda. *International Journal of Operations & Production Management*, *15*(4), 80–116. https://doi.org/10.1108/01443579510083622

Park, A., Nayyar, G., & Low, P. (2013). *Supply chain perspectives and issues: A literature review*. WTO and Fung Global Institute.

Pérez-Solà, C., Delgado-Segura, S., Navarro-Arribas, G., & Herrera-Joancomartí, J. (2019). Double-spending prevention for Bitcoin zero-confirmation transactions. *International Journal of Information Security*, *18*(4), 451–463. https://doi.org/10.1007/s10207-018-0422-4

Queiroz, M. M., & Fosso Wamba, S. (2019). Blockchain adoption challenges in supply chain: An empirical investigation of the main drivers in India and the USA. *International Journal of Information Management*, *46*, 70–82. https://doi.org/10.1016/j.ijinfomgt.2018.11.021

Queiroz, M. M., Telles, R., & Bonilla, S. H. (2019). Blockchain and supply chain management integration: A systematic review of the literature. *Supply Chain Management: An International Journal*, *25*(2), 241–254. https://doi.org/10.1108/SCM-03-2018-0143

Reason, J. (1990). *Human error*. Cambridge University Press.

Risius, M., & Spohrer, K. (2017). A blockchain research framework. *Business & Information Systems Engineering*, *59*(6), 385–409. https://doi.org/10.1007/s12599-017-0506-0

Saberi, S., Kouhizadeh, M., Sarkis, J., & Shen, L. (2019). Blockchain technology and its relationships to sustainable supply chain management. *International Journal of Production Research*, *57*(7), 2117–2135. https://doi.org/10.1080/00207543.2018.1533261

Sabz Ali Pour, F. (2021). *Application of a blockchain enabled model in disaster aids supply network resilience* [Dissertation, Old Dominion University Libraries]. https://doi.org/10.25777/FKR7-A212

Sanders, N. R., & Wagner, S. M. (2011). Multi-disciplinary and multi-method research for addressing contemporary supply chain challenges. *Journal of Business Logistics*, *32*(4), 317–323. https://doi.org/10.1111/j.0000-0000.2011.01027.x

Schwab, K., & Malleret, T. (2020). *COVID-19: The great reset*. World Economic Forum.

Sharawi, A., Sala-diakanda, S., Dalton, A., Quijada, S., Yousef, N., Rabelo, L., & Sepulveda, J. (2006). A distributed simulation approach for modeling and analyzing systems of systems. *Proceedings of the 2006 Winter Simulation Conference*, 1028–1035. https://doi.org/10.1109/WSC.2006.323191

Sjöberg, L. (1999). Risk perception in Western Europe. *Ambio*, *28*(6), 543–549.

Song, C. (2005). *A methodological framework for vulnerability assessment for critical infrastructure systems, hierarchical holographic vulnerability assessment (HHVA)* [Thesis, ETH Zürich]. https://www1.ethz.ch/lsa/education/arb/old/archive/da_song_05

Sweeney, J. L. (2002). *The California electricity crisis*. Hoover Institution Press.

Taleb, N. N. (2014). *Antifragile: Things that gain from disorder* (Reprint ed.). Random House Trade Paperbacks.

Terzi, S., & Cavalieri, S. (2004). Simulation in the supply chain context: A survey. *Computers in Industry*, *53*(1), 3–16. https://doi.org/10.1016/S0166-3615(03)00104-0

Tian, F. (2017). A supply chain traceability system for food safety based on HACCP, blockchain amp; Internet of things. *2017 International Conference on Service Systems and Service Management*, 1–6. https://doi.org/10.1109/ICSSSM.2017.7996119

Tierney, K. (2012). Disaster governance: Social, political, and economic dimensions. *Annual Review of Environment and Resources*, *37*, 341–363. https://doi.org/10.1146/ANNUREV-ENVIRON-020911-095618

Tokgoz, B. E., & Gheorghe, A. V. (2013). Resilience quantification and its application to a residential building subject to hurricane winds. *International Journal of Disaster Risk Science*, *4*(3), 105–114. https://doi.org/10.1007/s13753-013-0012-z

Tolmach, P., Li, Y., Lin, S.-W., Liu, Y., & Li, Z. (2021). A survey of smart contract formal specification and verification. *ACM Computing Surveys*, *54*(7), 148:1–148:38. https://doi.org/10.1145/3464421

Turner, B. L., Kasperson, R. E., Matson, P. A., McCarthy, J. J., Corell, R. W., Christensen, L., Eckley, N., Kasperson, J. X., Luers, A., Martello, M. L., Polsky, C., Pulsipher, A., & Schiller, A. (2003). A framework for vulnerability analysis in sustainability science. *Proceedings of the National Academy of Sciences of the United States of America*, *100*(14), 8074–8079. https://doi.org/10.1073/pnas.1231335100

Vamanu, B. I., Gheorghe, A. V., & Katina, P. F. (2016). *Critical infrastructures: Risk and vulnerability assessment in transportation of dangerous goods: Transportation by road and rail*. Springer International Publishing. www.springer.com/us/book/9783319309293

Wang, Y., Han, J. H., & Beynon-Davies, P. (2018). Understanding blockchain technology for future supply chains: A systematic literature review and research agenda. *Supply Chain Management: An International Journal, 24*(1), 62–84. https://doi.org/10.1108/SCM-03-2018-0148

Weber, E. U., & Hsee, C. (1998). Cross-cultural differences in risk perception, but cross-cultural similarities in attitudes towards perceived risk. *Management Science, 44*(9), 1205–1217.

WHO/EHA. (2002). *Disasters and emergencies: Definitions (No. 7656)*. Panafrican Emergency Training Centre. https://apps.who.int/disasters/repo/7656.pdf

Wieland, A., Handfield, R. B., & Durach, C. F. (2016). Mapping the landscape of future research themes in supply chain management. *Journal of Business Logistics, 37*(3), 205–212. https://doi.org/10.1111/jbl.12131

Yoo, M., & Won, Y. (2018). A study on the transparent price tracing system in supply chain management based on blockchain. *Sustainability, 10*(11), 4037. https://doi.org/10.3390/su10114037

2 Facets of Disaster Management

2.1 DYNAMICS OF DISASTER MANAGEMENT

The dynamic nature of disasters and response operations is an ideal place for volatility, uncertainty, complexity, and ambiguity in disaster supply chains; after all, such situations involve scarce information of all sorts: infrastructure, supplies, and requirements, especially in the initial phases. In fact, Richey (2009) claims that the main foundations of a disaster supply chain involve issues of collaboration, communication, and contingency planning. This might also be the case since different entities (including people) tend to pursue different goals and use different strategies in a disaster situation. Previous research suggests blockchain technology can be used to address a set of core challenges in disaster supply and logistics management (Apte & Petrovsky, 2016; Colicchia & Strozzi, 2012; Garcia & You, 2015; Kleindorfer & Saad, 2005):

- *Knowledge Management:* focusing on the lack of accurate and reliable data for analytics, excessive redundancy (and crosschecking), and abundancy (and independent) of data sources of different parties
- *Humanitarian Aids:* focusing on the lack of clear definitions of roles, coordination due to geographical dispersion, and systemic governance
- *Logistics Communication and Collaboration:* focusing on the analog gaps between victims and relief teams, lack of information sharing among all involved participants, and lack of an integrated global view concerning the dynamic of the disaster
- *Openness and Swift Trust:* focusing on the lack of traceability of the process flow and the limited visibility. Addressing such issues as how to obtain and distribute resources

However, there remains a need to investigate how the concept of disaster aid can be integrated with supply chain management, especially at the dawn of revolving complexity stemming from volatility, uncertainty, complexity, and ambiguity associated with disasters and the diversity of stakeholders, including victims.

2.2 DISASTER AID MANAGEMENT

We can all agree that disasters pose severe logistical challenges to emergency response due to the disruptions and the potential to create chaos (Holguín-Veras et al., 2007). Arguably, critical difficulties in responding to disasters involve collaboration and coordination. Hence, to design a reliable disaster supply aids system, knowledge about disaster supply chain operation and interaction and methods to analyze and coordinate the flows of priority goods are required (Holguín-Veras et al., 2007). If empowered by blockchain technology, such an approach would result in an integrated, more resilient, and effective platform for disaster aid management, thus alleviating limitations within the current disaster aid supply network.

Blockchain technology can provide an effective solution to disaster supply network resilience due to its three foundational tenets (Liang et al., 2017):

- The decentralized architecture of blockchain makes it robust against failures and attacks.
- Blockchains rely on a critical public structure, which allows the contents to be encrypted and thus challenging to crack. The architecture also records all operations permanently and yet transparently.
- Data in a blockchain is stored in a shared, dispersed, and fault-tolerant database for every member within the network. Moreover, blockchain technology can annul adversaries by harnessing the computational capabilities of the honest nodes, hence making information exchange resilient to manipulation.

At this point, suffice it to say that blockchains can enhance decision-making in disaster management by considering issues of transparency and trust in information management. However, one might be interested in the performance supply network in a disaster region. In this case, we assume that performance is directly related to the level of information being shared in real time. This affects the response time, meaning that delays can also be attributed to meeting provisions. We also assume that a blockchain-enabled model positively influences disaster complexity management and that such a model can be used to monitor supplies within a supply chain and logistics network. We suggest that the significance of blockchain technology should first be framed in the context of existing challenges in disaster supply chain and logistics management. It will then become clear how the features of blockchain can be used to address the identified gaps and challenges.

Disasters continue to inflict unwanted consequences around the globe, devastating results in human losses and infrastructure collapse (Cavdur et al., 2016; Tierney, 2012). A single disaster can cause losses of thousands of lives and billions of dollars—think of recent disasters; 2005 Hurricane Katrina (Katina, 2016), the 2010 Deepwater Horizon oil spill (Katina & Keating, 2022), and Japan's 2011 earthquake and tsunami (NOAA, 2021). In a disaster, delivering critical supplies is a challenging task because of possible damages to the infrastructures (physical and virtual) and transportation capacity limitations in the affected areas. Multidisciplinary research approaches are required to address the unique challenges of emergency logistics. Holguín-Veras et al. (2007) suggest that multidisciplinary approaches are necessary because disasters require a large volume of supplies, have a short time frame for a response, and contain a significant level of uncertainty about needs and availabilities in the affected area. The fundamentals of hazards, risk, disaster, and disaster supply chain are explored to set the stage for the need for supply chain and logistics management.

2.2.1 HAZARDS

A hazard can be defined as a source of danger that may or may not lead to an emergency or disaster (Hill, 2014). Each hazard carries an associated risk represented by the likelihood of the hazard and the consequences of that event. The realized hazard risk can produce an emergency event that is characterized as a condition displaying adverse effects (Haddow et al., 2017). Moreover, dealing with such consequences requires effort, especially in disaster management.

We arrive at the disaster zone when the conditions to respond to an emergency event exceed the emergency services capabilities. And at that zone, *hazard identification* becomes the basis of all emergency management activities. Any knowledge and insights derived from the identified hazards serve as the groundwork for disaster management preparedness and mitigation actions (Haddow et al., 2017). Hazard identification aims to establish an exhaustive list of hazards for analysis. While hazards can vary based on the circumstances, a general set of categories for disaster hazards includes natural (e.g., coastal erosion, drought, earthquakes, extreme temperatures, flood, hail, hurricanes, mass movements, severe winter storms, storm surges, thunderstorms, tornadoes, tsunamis, volcanic eruptions, wildfires); technological (e.g., dam failures, hazardous materials incidents, nuclear accidents, structural fires, terrorism); chemical, biological, radiological, and nuclear. The last three will typically involve chemical, biological, radiological, and nuclear devices and incidents.

The magnitude and existence of hazards should be identified via historical and scientific data interpretation using multiple sensing devices and models. Communication technologies are crucial in hazard-specific detection and monitoring (Samarajiva, 2005).

2.2.2 RISK MANAGEMENT

The process of evaluating the identified risks and designing strategies and procedures to mitigate the risk factors is the management of risk. And since each hazard carries an associated risk represented by the likelihood of the hazard and the consequences of that event, risk management can be seen as hazard management. Risk management involves four phases: hazard identification, risk assessment, risk analysis, and risk treatment (NASA, 2007). Hazard identification involves checklists of potential risks and evaluating the likelihood that those events might happen on the project. Hazard assessment includes both the identification of potential risk and the evaluation of the potential impact of the hazard. Risk analysis provides an in-depth analysis of the situation. Risk treatment is more of a mitigation plan, which looks at the plan to reduce the impact of the just-in-case events, with the goal of risk avoidance, risk sharing, risk reduction, and risk transfer.

2.2.3 DISASTER

There are several definitions of disaster. For example, according to the Center for Research on the Epidemiology of Disasters, a disaster can overwhelm the local capacity, requiring the national or international level for external support; such an event is often unforeseen and sudden and causes destruction, human suffering, and significant damage. Similarly, the World Health Organization describes a disaster as any manifestation that causes damage, destruction, deterioration of health, ecological disruption, human suffering, loss of human life, and health services on a scale sufficient to warrant an astonishing response from outside the affected community (Haghani & Afshar, 2009). Additionally, the American Red Cross defines a disaster as an occurrence (or a situation) that produces human suffering to the extent that victims cannot improve on their own without support (Haghani & Afshar, 2009). Moreover, the International Strategy for Disaster Reduction suggests that a disaster is a "severe disruption of the functioning of society, posing a significant, widespread threat to human life, health, property, or the environment whether caused by accident, nature, or human activity, and whether developing suddenly or as a result of complex, long-term processes" (Day et al., 2012, p. 24).

Furthermore, the Federal Emergency Management Agency (FEMA) defines a disaster as a natural catastrophe, technological accident, or human-caused event resulting in severe property damage, deaths, or multiple injuries. Typically, a large-scale disaster exceeds the response capability of the local jurisdiction and requires state and potentially federal involvement (FEMA, 1996). However, the official position of the United States is that a disaster is "any natural catastrophe (including hurricane, tornado, storm, high water, wind-driven water, tidal wave, tsunami, earthquake, volcanic eruption, landslide, mudslide, snowstorm, or drought) or, regardless of cause, any fire, flood, or explosion, in any part of the United States, which is the determination of the President causes the damage of sufficient severity and magnitude to warrant major disaster assistance under this Act to supplement the efforts and available resources of States, local governments, and disaster relief organizations, in alleviating the damage, loss, hardship, or suffering caused thereby" (Robert T. Stafford Disaster Relief and Emergency Assistance Act, PL 100–707, Signed into Law November 23, 1988; Amended the Disaster Relief Act of 1974, PL 93–288, 2001) (FEMA, 2001).

Regardless of one take on disaster definition, many of these disasters cannot be stopped by humans. The next best thing might be resilient-based approaches to help with management, mitigation, and recovery from disasters, especially in the wake of the frequency of disasters and their impact. Table 2.1 shows the total number of people affected by natural disasters (Haghani & Afshar, 2009).

Moreover, Figure 2.1 shows the number of disastrous events from 1970 to 2019 (Ritchie & Roser, 2014). For example, in 2018, SwissRe Institute estimated that more than 11,000 people were victims of disasters. The economic losses were estimated to be $155 billion.

Disasters can be categorized into three types: natural, man-made, and hybrid disasters. Natural disasters are large-scale geological or meteorological events that have the potential to cause loss of life or property. These typically include tornadoes and severe storms, hurricanes and tropical storms, floods, wildfires, earthquakes, and drought. Man-made disasters are those that are generated by human actions. Examples include acts of terrorism, incidents of mass violence, industrial accidents, and shootings. Hybrid disasters involve human error and natural forces such as deforestation, leading to erosion after a tropical storm (Shaluf, 2007).

TABLE 2.1

Number of People Affected by Natural Disasters between 2000 and 2018

Event	Average (2000–2017)	2018
Drought	58,734,128	9,368,345
Earthquake	6,783,729	1,517,138
Extreme temperature	6,368,470	396,798
Flood	86,696,923	35,385,178
Landslide	263,831	54,908
Mass movement (dry)	286	0
Storm	34,083,106	12,884,845
Volcanic activity	169,308	1,908,770
Wildfire	19,243	256,635
Total	**193,312,310**	**61,772,617**

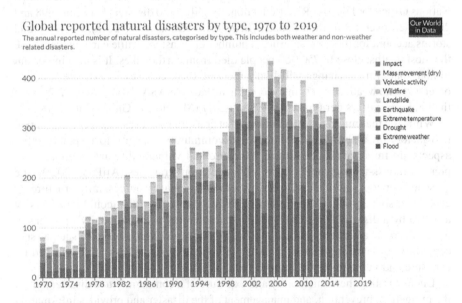

Global reported natural disasters by type, 1970 to 2019
The annual reported number of natural disasters, categorised by type. This includes both weather and non-weather related disasters.

Our World in Data

■ Impact
■ Mass movement (dry)
■ Volcanic activity
■ Wildfire
■ Landslide
■ Earthquake
■ Extreme temperature
■ Drought
■ Extreme weather
■ Flood

FIGURE 2.1   Number of global natural disasters by type from 1970–2019.

Alarming is the increasing rate of occurrence and impact of natural disasters. In 2021, for example, insured losses from natural disasters again exceeded the previous ten-year average, continuing the trend of an annual 5%–6% rise in losses seen in recent decades (Swiss Re, 2021). It seems to have become the norm that at least one secondary perilous event, such as severe flooding, winter storm, or wildfire, each year results in more than USD 10 billion in losses. At the same time, Hurricane Ida is a stark reminder of the threat and loss potential of peak perils. Just one such event hitting densely populated areas can strongly impact the annual losses (Swiss Re, 2021).

FIGURE 2.2   The number of disasters by type from 1998 to 2017.

Natural disasters affect millions of people every year. According to the United Nations Office for Disaster Risk Reduction, roughly 7250 disasters between 1998 and 2017 killed over 1.3 million people. (Mizutori & Guha-Sapir, 2018). Flooding and storms account for the most significant number of disasters, while earthquakes cause the most deaths; close to 750,000 people died from earthquakes. It is also important to note that climate change will likely increase the frequency of extreme heat and other extreme weather events in the coming decades (NASA, 2001). Figure 2.2 shows the total of disasters per type from 1998 to 2017 (Mizutori & Guha-Sapir, 2018).

While the maximum for establishing what constitutes a disaster is dependent upon two factors (availability of resources and community capacity to respond), other aspects are important. For example, exposure and vulnerability are essential components in risk-management efforts and adaptation strategies (AliPour, 2021). The presence of people, animals, ecosystems, environmental resources, infrastructure, or economic, social, and cultural assets in places and settings that could be adversely affected by a disaster is called exposure. Vulnerability is a community's propensity to be adversely affected by a disaster, taking into consideration factors such as susceptibility to harm and lack of capacity to cope and adapt. Risk is determined by exposure and vulnerability to hazards (NASA, 2001).

Understanding the vulnerability and exposure of a community to a disaster aids in the mitigation, prevention, and management of the disaster and provides information to help with response and relief efforts.

2.2.4 Disaster Management

Disaster management is the process that enables the possibility to prepare, respond, and recover from disasters as well as minimize destruction. Disaster management is related to the discipline of emergency management, which involves preparing for disasters before, during, and after a disaster. However, it also involves supporting and rebuilding a society long after a disaster (Haghani & Afshar, 2009). Emergency management has deep roots in ancient history. In fact, it has been suggested that the account of Moses parting the Red Sea to control floods could be considered the first emergency management attempt (Haddow et al., 2017). Moreover, emergency management is integral to daily activities that we integrate into daily decisions without experiencing a disaster. During the 1950s, the Cold War era, the term "emergency management" became widely used. The creation of the Office of Emergency Preparedness by the White House further popularized emergency management in 1961 to deal with natural disasters more effectively. The Federal Emergency Management Agency (FEMA) was established in 1978.

To reduce the conflicts facing FEMA, the Integrated Emergency Management System (IEMS) was created with the idea of capturing all kinds of hazards along with aspects of control, direction, and warning as fundamental functions of all sorts of disasters. The emergency management system had a major evolution during the 1990s. FEMA launched a national initiative called Building-Resistant Communities to promote a new community-based approach (Tokgoz, 2012). In this new approach, all the stakeholders, including the business sector, were involved in identifying risks and establishing plans to reduce them. This approach would promote a better life quality, protect and enhance natural resources, and create sustainable economic development for citizens (Haddow et al., 2017). The strategic goals of FEMA revolve around the following (Haghani & Afshar, 2009):

- Building community trust and confidence via performance and stewardship
- Delivering easily accessible and coordinated assistance for all programs
- Empowering people so that people can support FEMA's missions (i.e., provide reliable information at the right time to users and strengthen the ability of the homeland to address disasters, emergencies, and terrorist attacks)

However, there isn't one global formula for disaster management capacity development. In fact, most nations have their own legal frameworks that guide disaster management. In the end, however, as a process, disaster management involves developing and implementing policies that are concerned with four main phases:

- *Mitigation:* measures to reduce risks to people and property from the hazards and their effects. The essential part of any mitigation strategy is analyzing all the potential hazards in a particular area (Haddow et al., 2017; Petak, 1985). Mitigation involves deciding actions to be undertaken to reduce health, safety, and welfare risks, including the implementation of risk-reduction programs.
- *Preparedness:* a readiness state to respond to a disaster or any other type of emergency. The heart of preparedness is planning, training, and exercising to develop the capacity to respond and recover from emergencies and disasters. Preparedness is preparing a response plan and training to save lives and reduce disaster damage, including identifying critical resources and developing necessary agreements among responding agencies (Petak, 1985).
- *Response:* measures to save lives, protect property, and meet basic human needs. Responding to disaster events is the most visible activity of emergency management. Clear lines of communication and coordination of numerous

agencies are required to have a successful response, a strong command, and a control system (Haddow et al., 2017; Petak, 1985).

- *Recovery:* related to choices and action relative to permanently repairing, rebuilding homes, rebuilding infrastructures, replacing properties, restoring businesses, and resuming employment. The recovery process may require short- and long-term goals to be defined to reduce susceptibilities and vulnerabilities. An active recovery aims to bring all the players together to plan, finance, and implement a recovery strategy that will rebuild the affected area safer and more secure as quickly as possible (Haddow et al., 2017; Petak, 1985).

Completing the phases of disaster management, the national preparedness directorate of FEMA developed a systematic approach toward a cyclical process to establish and improve preparedness (Haddow et al., 2017), involving five constant steps as suggested in Figure 2.3:

- *Step 1:* involves preparing a plan and includes planning to protect the citizens, property, and essential services after a disaster and ensuring the viability of the community is sustained despite the existing hazard risks.
- *Step 2:* involves acquiring equipment, and yet actual possession and access to the required equipment (e.g., communications equipment, personnel protective equipment, proper search and rescue equipment) can be limiting factors in this phase.

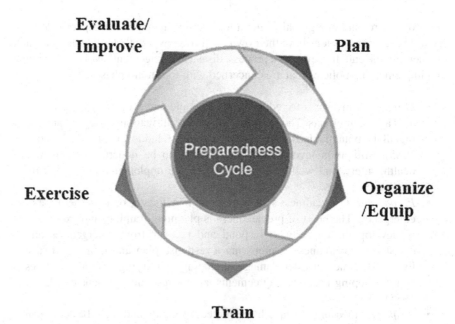

FIGURE 2.3 FEMA's preparedness planning cycle.

- *Step 3:* involves training based on the developed plan. All the key stakeholders operating in the community are required to participate in emergency response training.
- *Step 4:* involves exercising the developed plan. An improved understanding of response realities can be achieved via different exercises and simulations. Also, the shortfalls or failures in the previous steps can be identified.
- *Step 5:* involves evaluation and improvement based on lessons learned. Lessons learned are documented to be applied for the subsequent simulations. This step is the product of two sources: (i) exercises in Step 4 and (ii) results from experiencing actual disasters.

While it is easy to see that efficient disaster preparedness involves the capacity to receive and convert the information (i.e., hazard detection and monitoring system) into accurate, credible, and timely warning/alert, it is equally critical that media is used to distribute warnings to different stakeholders (Samarajiva, 2005). And in this case, the Incident Command System established a set of planning and management systems that can help with systematic coordination during disasters.

The Incident Command System (ICS) is a highly standardized, top-down, military-based management structure that supersedes every other agency responding to an event. This management tool is used to meet the requirements of small or large emergency and non-emergency situations. It represents best practices and has become the standard for emergency management across the United States. ICS standardizes position titles, responsibilities, and terminology. It uses titles that aren't necessarily used in regular jobs, and it's very clear about what responsibilities are. The system's benefits are that it establishes common processes for planning and managing resources and allows for integrating facilities, equipment, personnel, procedures, and communications within a typical organizational structure.

In a significant crisis (e.g., Hurricane Katrina), the whole structure could be implemented, unlike minor incidents where fewer positions might be filled. One person can fill more than one role. The rule of thumb is that the role above fills every role that is left vacant. The incident commander is always the first position created (and sometimes the only position created), and it's always the last position demobilized. Generally, an incident commander will appoint a command staff and general staff. These three command staff positions are meant to shield the incident commander from getting too much information or getting too bogged down with little decisions so that they can focus on the overall response instead. The command staff positions include public information officer (PIO), safety officer, and liaison officer.

The PIO is responsible for the following:

- Advising the incident commander on information dissemination and media relations (note that the incident commander approves information that the PIO releases).
- Obtaining information from and providing information to the planning section.
- Obtaining information from and providing information to the community and media.

The safety officer is responsible for the following:

- Advising the incident commander on issues regarding incident safety.
- Working with the operations section to ensure the safety of field personnel.
- Ensuring the safety of all incident personnel.

The liaison officer is responsible for the following:

- Assisting the incident commander by serving as a point of contact for representatives from other response organizations.
- Briefing and answering questions from supporting organizations.

The general staff falls into four different sections. Each section is responsible for a central functional area of the incident. Depending on the size of the disaster, there is typically one section chief who can appoint more staff below them. The general staff sections include operations, planning, logistics, and finance. The operations section is the first (and sometimes only) section created. It directs and coordinates all tactical incident operations.

The planning section collects, evaluates, and displays incident intelligence and information. It prepares and documents incident action plans. It also tracks resources assigned to the incident and maintains incident documentation. It is also responsible for developing plans for demobilization.

The logistics section orders obtains, maintains, and accounts for essential personnel, equipment, and supplies. It provides communication planning and resources. The finance/administration section negotiates and monitors contracts, performs timekeeping, analyzes cost, arranges compensation for injury or damage to property, and documents reimbursement such as memorandums of understanding (CCAHA, 2020).

In the recovery phase, key factors include identifying needs and resources, providing accessible housing, addressing the care and treatment of affected persons, informing residents (and preventing unrealistic expectations), implementing additional measures for community restoration, and incorporating mitigation measures and techniques.

In developing a recovery plan, decision-makers should consider many factors (e.g., building codes, finances, land-use planning techniques, oversight, zoning, etc.). A long-term recovery plan aims to identify the impacts of a disaster on the area and develop a process to ensure the recovery of the whole community, not just a select few (Haddow et al., 2017). These activities are instrumental in identifying vulnerable areas, accelerating funding for rebuilding the post-disaster environment, addressing regulatory and environmental requirements for rebuilding, and minimizing the economic and social disruption.

However, there is a scarcity of literature discussing the topic of catastrophe recovery (Chang, 2010). The key concerns in this phase include the operational problems of damage assessment and cleanup and the key characteristics of food and monetary aid collection, allocation, and distribution. Table 2.2 is based on Altay and Green (2006) and comprehensively describes the emergency in terms of four programmatic phases.

TABLE 2.2

Operational Activities Associated with Disaster Management Phases

Phase	Phase Description	Operational Activities
Preparedness	Determining activities and procedures to prepare the community for responding to a disaster	• Recruiting personnel for the emergency services and community volunteer groups • Emergency planning • Development of mutual aid agreements and memorandums of understanding • Training for both response personnel and concerned citizens • Threat-based public education • Budgeting for and acquiring vehicles and equipment • Maintaining emergency supplies • Construction of an emergency operations center • Development of communications systems • Conducting disaster exercises to train personnel and test capabilities
Response	Using supplies and procedures to preserve life, property, environment, and social, economic, political structures	• Activating the emergency operations plan • Activating the emergency operations center • Evacuation of threatened populations • Opening of shelters and provision of mass care • Emergency rescue and medical care • Firefighting • Urban search and rescue • Emergency infrastructure protection and recovery of lifeline services • Fatality management
Recovery	Specifying long-term actions after the immediate impact of the disaster to stabilize the community and restore some the aspect of normalcy	• Disaster debris cleanup • Financial assistance to individuals and governments • Rebuilding of roads and bridges and critical facilities • Sustained mass care for displaced human and animal populations • Reburial of displaced human remains • Complete restoration of lifeline services • Mental health and pastoral care
Mitigation	Employing measures to prevent hazards or decrease their impacts in case of occurrence	• Zoning and land use controls to prevent the occupation of high-hazard areas • Barrier construction to deflect disaster forces • Active preventive measures to control developing situations • Building codes to improve disaster resistance of structures • Tax incentives or disincentives • Controls on rebuilding after events • Risk analysis to measure the potential for extreme hazards • Insurance to reduce the financial impact of disasters

Obviously, governments may take the lead in dealing with each phase of a disaster. However, to have a practical mitigation strategy, community involvement is necessary. And beyond community involvement, there is a need for technical knowledge that may help define historical occurrences, public education, and media attention—all of which are necessary to recognize each vulnerability. To enhance the mitigation, the government can offer political visions, enticements, and fines for not taking action. Moreover, a community partnership should be built, and the steps would be communicated to all the stakeholders. The preparedness phase should not only focus on protecting the citizens, property, and essential services after a disaster but also ensure sustainability despite the existing hazard risks. The planning phase can be used to understand disaster vulnerabilities. Vulnerability assessment can be the basis for assessment of preparedness levels as well as determining needed capabilities—resources and otherwise—in case of an emergency.

All countries face risks and hazards and, from time to time, will be confronted with disasters. Therefore an essential aspect of addressing disasters might not be when or if the disaster will occur but rather response capacity. Response capacity is the ability to deal with a disaster effectively and is linked to several factors: propensity for disaster, local and regional economic resources, government structure, and technological, academic, and human resources (Haddow et al., 2017). Moreover, the response may vary based on the size of a disaster. For example, a large-scale disaster may require an international response (and assistance) beyond the capacity of a single nation. As the response capacity of a nation's emergency management structure is overwhelmed by the severity of hazards and consequences, the event could become a global disaster, as in the case of COVID-19. Coppola (2006) posits that international disaster management stakeholders (i.e., participants) will include governments of the affected nations, international financial institutions, international organizations, local and regional donors, local first responders, nonprofit organizations, private organizations, regional organizations, and victims.

International disasters pose significant challenges in coordinating and collaborating. The United Nations' Inter-Agency Standing Committee, established in 1992, serves as a platform within which a broad range of humanitarian partners may come together to address humanitarian needs and is the basis for (i) developing and agreeing on system-wide humanitarian policies, (ii) allocating responsibilities among agencies in humanitarian programs, (iii) developing and agreeing on a shared ethical framework for all humanitarian activities, and (iv) identifying areas where gaps in mandates or lack of operational capacity exist. However, no global stock positioning system offers information regarding quantity, quality, geographical location, and ownership of stocks. Nonetheless, Balcik and Beamon (2008) provide an analytical approach to making effective and efficient facility location and stock pre-positioning decisions to assist decision-makers. Much research remains in developing systematic approaches supporting infrastructure for relief chain design, especially in facilitating response capacity and performance. Figure 2.4 depicts the flow of resources in disaster relief, as Balcik and Beamon (2008) suggest.

It can be expected that several organizations would be involved in different tasks and responsibilities in disaster management. The challenge is how such organizations can work together as effectively as possible (Scholtens, 2008). However, a lack

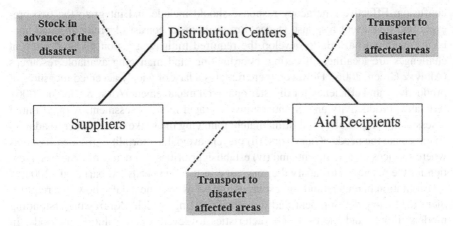

FIGURE 2.4 The flow of resources for disaster relief.

of knowledge can act as a deterrence for strong collaboration. In such cases, non-governmental agencies are invaluable support. Nongovernmental agencies can act as information-gathering bodies and are crucial in establishing accurate assessments of damages and needs. They can provide specific technical skills to allow reaching a larger population with a more excellent capability in less time. Moreover, their fund-raising abilities can bring more significant cash to address the victims' needs. For example, in the 2004 Asian tsunami, more than 40 countries and 700 NGOs assisted (Balcik et al., 2010).

Disaster management literature is replete with models, guidelines, and procedures for more effective disaster plans while claiming that traditional approaches are useful in disaster management from a single government's perspective (Luna & Pennock, 2018). Much of the research in the disaster management field is targeted at public servants, government agencies, and insurance firms charged with responding in times of crisis (Hale & Moberg, 2005). As one would expect, making a decision in a disaster is stressful. One has to be able to deal with complexity, unpredictability, and the dynamics of the emergency while effectively navigating all sorts of pressures—personal danger, the sight of causalities, time pressure, and communication. Paton and Flin (1999) examined the stress in decision-making in response to a disaster categorized in environmental, organizational, and operational sources. Different tools and approaches have been applied to enhance the decision-making process in case of a disaster. For example, Gunes and Kovel (2000) built a database in a geographic information system framework to help integrate emergency management into decision-making. Sahebjamnia et al. (2017) developed a hybrid decision support system framework to respond to large-scale disasters consisting of a knowledge base and a ruled-based simulator. A rule-based model comprises a set of rules, which can be processed by general-purpose simulation and analysis tools to achieve different objectives (Chylek et al., 2014).

Disaster operation management is also multi-organizational by nature. However, the organizations are loosely connected, leading to authority ambiguity and management confusion. Effective emergency response should include and integrate the two pre-event and post-event response stages. Pre-event tasks consist of predicting and analyzing potential hazards to develop the required mitigation action plans. Post-event challenges are locating, allocating, coordinating, and managing available resources (Altay & Green, 2006). However, there remains a lack of widely accepted measures of productivity and efficiency for disaster operation management (Altay & Green, 2006). Arguably, productivity and efficiency involve rapid initial assessment. A rapid initial assessment is especially vital immediately following the strike of a disaster to address (i) anticipate the needs of survivors, (ii) prevent avoidable mortality, (iii) identify areas where services were deficient, and (iv) establish priorities and determine the best location for the resources to satisfy the survivors' immediate needs (Akbari et al., 2004).

Limitation in understanding resources and types of critical data shows the requirement for theory development and hypothesis testing, which requires understanding models' inputs and the event characteristics to develop new solution methods. In disaster management, it should be recognized whether social interactions are helping or hindering people in reducing their vulnerability to hazards (Haddow et al., 2017). Due to the nature of disasters, multiple agencies need to work together to monitor the response and manage many people responding to the affected area. All agencies must have an integrated system to operate under one overall response management system. Duplication of efforts, lack of coordination, and communication problems hinder the involved parties from responding to a disaster (Haddow et al., 2017).

The first step in any emergency response is to assess the extent and impact of the damage caused by the disaster and the capacity of the affected population to meet its immediate survival needs (degree of vulnerability). Although the impact may vary considerably from one disaster to another, based on the International Federation of Red Crescent Societies (IFRC), specific needs include food, shelter, essential items, medical care, sanitation, waste disposal, and psychosocial support. However, public-sector problems are not well-defined. Moreover, such problems include high-societal issues overlaid with political overtone. These issues should focus on new networks and organizational structures, especially those developed to facilitate disaster communication and coordination (Altay & Green, 2006).

2.3 COMPLEXITY IN DISASTER MANAGEMENT

Again, disaster management is complex and multifaceted (Fogli et al., 2017). The disaster management domain faces an increase in complexity and a decrease in the predictability of operational scenarios (Gunes & Kovel, 2000). The newly emerging complexity science enables a detailed understanding of the likelihood of disasters (Casti, 2012). The essential purpose of unraveling the complexities of disaster management is to recognize the interdependencies between response agencies and the other involved agencies and how they intersect to promote resilience before, during, and after a disaster. It is now possible to empirically explore the complexity of disasters to determine levers for action where interventions can be executed to facilitate the collaborative effort (O'Sullivan et al., 2013). Despite these strides, there remain limitations; strangely, these limitations are associated with understanding behaviors of entities you would think we understand better, namely organizational, institutional, and human behavior (Gerber, 2007). A vital aspect of this complexity is communication and collaboration among the aids network participants and providing real-time information in disasters.

2.3.1 COMMUNICATION IN DISASTER MANAGEMENT

The United Nations has a leadership role and a prescribed practice in a major international disaster. Still, the role of the government must be recognized since all the global actions are based on their request. The participants must be committed to identifying the needs and accept to appeal to the affected area. Due to the sheer number of responding agencies, a vital and immediate component is required for coordination. As one would expect, each entity may have a specific skill/service they offer. And as such, not only is coordination imperative, but also, smooth coordination and cooperation can lead to the tremendous success of the disaster response. The state's sovereignty should be based on recognizing political authority characterized by territory and autonomy. For equality in relief distribution, the challenge is that sometimes a particular group of aid needs is favored over others. These biases can be categorized into two: discrimination (e.g., gender bias) and class bias. For capacity building and linking relief with development, disasters present an opportunity to rebuild old, ineffective structures and develop policy and practice toward a more resilient community. Linking relief and development should not be a deviation from the agencies' missions. Training and information exchanges are essential in the reconstruction phase to mitigate repeat disasters and increase the likelihood of a nation being developed sustainably.

Information technologies can support communication challenges throughout a disaster response by focusing on interactions between agencies with multiple developers or vendors operating different software. For example, Papadopoulos et al. (2017) provide seven categories: infrastructure development companies (6); transporters (30); warehousing (18); army logistics (2); border road organization (1); nongovernmental organizations (30); and medical aid agencies (30) as the main types of organizations involved in disaster response (the bracketed numbers are approximations used in the development of the network). To facilitate communication among the entire network, Imran et al. (2015) suggest the use of machine-understandable ontologies to define, categorize, and maintain a relationship between several concepts to be able to facilitate mutual understanding and integrated communication.

Clearly and in such a case, public information should be coordinated and integrated across all the involved parties. Communication would be an essential function in such an emergency situation. The information communication regarding preparedness, prevention, and mitigation can promote actions that reduce the risk of future disasters. Therefore, communication among participating agencies is imperative to share information as well as accomplish respective responsibilities effectively.

Additionally, an efficient disaster management operation requires goals, policies, as well as prioritized communication among stakeholders. Manoj and Baker (2007) identified three main communication challenges in the disaster responding system: technological, sociological, and organizational. These are critical factors in developing and maintaining a healthy and effective disaster communication system. The responding organizations make interagency communication. Without appropriate communication, decision-makers cannot gain an accurate picture of events to make decisions or execute them properly (Cartier et al., 2009). Good quality communication is essential for ensuring that appropriate information is available and delivered promptly (Paton & Flin, 1999). Cooperation (Hu et al., 2018) and interagency coordination (Smith & Dowell, 2000) are necessary to achieve the desired results in all phases of disaster management. Moreover, effective collaboration and leadership strategy are also necessary, especially in catastrophic situations (Waugh & Streib, 2006). The critical factors that describe communications in planning are channels of communication, the sources of the communication, and message distribution or delivery strategy.

Furthermore, in dealing with disaster risk communication, it is necessary to understand the situation and work with the media (Glik, 2007). Haddow et al. (2017) suggest that the disaster communication system must include (i) a definition of an effective disaster communications strategy, (i) an examination of media (past and present) for sharing information, and (iii) a specification for effective disaster communication in the future.

As one would suspect, communication must be involved in all four phases of emergency management (Haddow et al., 2017):

- *Mitigation:* promote implementation strategies, technologies, and actions that will reduce the loss of lives and property.
- *Preparedness:* communicating preparedness messages to encourage and educate the public in anticipation of disaster events.
- *Response:* providing notification, warning, evacuation, and situation report on an ongoing disaster.
- *Recovery:* providing registration information and receiving disaster relief to the affected individuals and communities.

At this point, suffice it to say that a communication specialist is needed within the emergency management team for communication as well as coordination. Two types of coordination (i.e., vertical and horizontal) are necessary within a relief chain. Vertical coordination is about the coordination of upstream and downstream activities, while horizontal coordination occurs among the organizations at the same level (Gerber, 2007). While many are involved in communication, the primary audiences for emergency management communication tend to involve the general public, disaster victims, business community, media outlets, elected officials, community officials, first responders, and volunteer groups.

Haddow et al. (2017) posit that to have effective communication in a disaster, the following are necessary:

- *Communication Plan:* protocols for collecting information from various sources, analyzing data to identify the needs and match the available resources to those needs and disseminating information about current conditions and actions to the public via media are all part of planning. Moreover, the response phase includes a plan for protocols to monitor the media, identify new sources, and evaluate the effectiveness of disaster communications. The recovery phase should include plans on available resources to help rebuild the disaster region.

- *Incoming information:* all possible sources of information should be identified, and working relationships be developed during non-disaster periods.
- *Messengers:* the authority figure that informs the public and media decisions. It should be determined what types of information will be delivered by which messenger prior to a disaster.
- *Monitor, update, and adapt:* staff should routinely monitor the media outlets to identify problems and issues early in the process and shape communication strategies to address them. The regular monitor helps in identifying rumors and misinformation and speed corrections. Monitoring information can be used to update communication plans, strategies, and tactics.
- *Outgoing information:* discover the best media to deliver preparedness and hazard mitigation messages and communicate with targeted audiences in response and recovery phases.
- *Staffing:* the involved organizations in emergency management should establish an ongoing communication staff capability. Staff will be required to establish and maintain working relationships with media sources.
- *Training and exercises:* the staff and messengers should be well trained to have effective disaster communication operations. The training includes media relations, new media, and marketing.

In the disaster response phase, the organizational structure for disaster management has been divided into three levels of coordination to shape multidisciplinary collaboration (Scholtens, 2008): administrative coordination, operational leadership, and presence (and operations) at the place of the disaster.

New technologies have changed the collaboration and distribution methods. These technologies modified the flow of information, replaced the centralized top-down model, and created ad hoc distributive information networks. Risk communications that advocate for disaster preparation should be imminent to motivate behavior change successfully. Inconsistency happens when the event is evolving, but the information is not updated regularly (Glik, 2007). The classical communication model of source, channel, message, receiver, effect, and feedback is the basis for transmitting social information. During hurricane Katrina, the communication capabilities became so weak that officials had to use human couriers to transmit messages (Majchrzak et al., 2007). The lack of interaction and institutional overlap among involved communities generate conflicts (Schipper & Pelling, 2006). Therefore, there remains a need for platforms that can facilitate collaboration among the entire network and in real time.

2.3.2 DISASTER GOVERNANCE

It is not feasible for one entity to command compliance among all participants. Since a disaster is a complex social problem, it cannot be expected to fit precisely within the space of a single entity. In such cases, management (and governance) must rely on the development and diffusion of many norms, state regulations and self-regulation, mechanisms of the market, and other processes (negotiation, participation, and engagement) that facilitate collective decision-making and activities (Keating et al., 2022). Governance via networks of collaboration and diverse entities yields to address the issues due to flexibility, adaptability, and mobilization capabilities of various resources. In this case, disaster governance is "the interrelated sets of norms, organizational and institutional actors, and practices that are designed to reduce the impacts and losses associated with disasters" (Tierney, 2012, p. 344). Since governments are not on a global scale, transnational governance processes are required to provide coordination mechanisms in the case of a disaster. Many disaster risk reduction attempts are dependent on cross-border collaboration and complex governance arrangements. Therefore, disaster governance is tied to risk, environmental, and earth governance (Tierney, 2012).

Decision-making and communications are centralized under the traditional approach (Luna & Pennock, 2018). Furthermore, Birkmann and von Teichman (2010) suggest that disaster challenges can be broken down into issues regarding scales, knowledge, and norms in a centralized knowledge database. Applying a central command and control system within a dynamic organization will lead to insurmountable restrictions (Scholtens, 2008). Providing an optimized decision-making process is impossible because a full range of information is never available in a dynamic disaster management organization. In such conditions, a distributed decision-making approach will be more efficient within complex and dynamic organizations (Scholtens, 2008). Examining FEMA's emergency management approach suggests that organizations collaborating to mitigate are more likely to be successful (Luna & Pennock, 2018). Still, there remains a lack of a *decentralized network* approach to disaster supply management supported by blockchain technology.

2.3.3 SUPPLY CHAIN

Disaster management has many supportive elements, including the supply chain. The supply chain is an integrated process in which various business entities work together to acquire raw materials, convert them to specified final products, and deliver them to retailers. The chain is characterized by a forward flow of materials and a backward flow of information (Beamon, 1998). The history of the supply chain initiative included traces of the textile industry with the quick response program. Later, the grocery industry used efficient consumer response. A supply chain analysis was commissioned by Kurt Salmon Associates in 1985. The outcomes of the study illustrated a long delay in warehouse or transit, which resulted in significant losses to the industry by the inventory costs and lack of the right product in the right place at the right time (Lummus & Vokurka, 1999). A supply chain is comprised of two fundamental and integrated phases (i) production planning and inventory control and (ii) distribution and logistics (Beamon, 1998). Figure 2.5 elaborates on the two integrated phases supply chain.

Interest in supply chain management has steadily increased since the 1980s as managers seek to capitalize on potential benefits that can accrue due to collaborative efforts within and beyond their organizations. However, this should not be taken

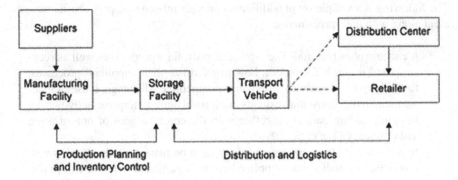

FIGURE 2.5   Elements of the supply chain process.

as an indication of uniformity across all disciplines. In fact, several definitions of supply chain management (and the associated activities) exist in different fields. The Supply Chain Council suggests that supply chain management involves "every effort involved in producing and delivering a final product from the supplier's supplier to the customer's customer. Four basic phases (i.e., plan, source, make, and deliver) define these efforts. These efforts also address managing supply and demand, sourcing raw materials, manufacturing and assembly, warehousing and inventory tracking, order entry and order management, distribution across all channels, and delivery to the customer (Lummus & Vokurka, 1999). Lummus and Vokurka's research (1999) offers several definitions of supply chain management. However, the summation of the matter is that supply chain management involves "all activities involved in delivering a product from raw material through to the customer including sourcing raw materials and parts, manufacturing and assembly, warehousing and inventory tracking, order entry and order management, distribution across all channels, delivery to the customer, and the information systems necessary to monitor all of the activities."

The sphere of the supply chain can also be complex. For example, Harland (1996) suggests that in managing supply chains, one must consider business activities and relationships in contexts of (i) internally within an organization, (ii) with immediate suppliers, (iii) with first- and second-tier suppliers and customers along the supply chain, and (iv) with the entire supply chain.

Given the aforementioned, one can rightfully conclude that supply chain management aims to identify the most important sets of outcomes and their articulation. The articulation is achieved through metrics and performance measurement systems. Then systems need to be designed and implemented that make the outcomes inevitable (Day et al., 2012). Additionally, supply chains are dynamic due to changes in the environment, technology, customer, and corporate strategy. Furthermore, supply chain management starts with the extraction of raw materials through the manufacturers, wholesalers, retailers, and the final users and in some cases includes recycling or reusing the products or materials (Haghani & Afshar, 2009). Noteworthily, the field contains more empirical-based studies than theoretical work. As such, less value seems to be placed on the field's theoretical foundations (Croom et al., 2000).

The following is a sample set of additional concepts related to supply chain management along with their proponents:

- It encompasses materials (i.e., raw materials, final product) as well as recycling and reuse. It focuses on how firms utilize their suppliers' processes, technology, and capability to enhance competitive advantage. It uses a management philosophy that extends traditional intra-enterprise activities by bringing trading partners together with the common goal of optimization and efficiency (Tan et al., 1998).
- It aims to build trust, exchange information on market needs, develop new products, and reduce the supplier base to a particular original equipment manufacturer to release management resources for developing meaningful, long-term relationships (Berry et al., 1994).
- An integrative approach to planning and controlling the materials flow from suppliers to end users (Jones & Riley, 1985).
- The external chain is the whole exchange chain from raw material source to the various firms involved in extracting and processing raw materials, manufacturing, assembling, distributing, and retailing to ultimate end customers (Saunders, 1995).
- A network of firms interacting to deliver products or services to the end customer, linking flows from raw material supply to final delivery (Ellram, 1991).
- A network of organizations is involved through upstream and downstream linkages in the different processes and activities that produce value in the form of products and services in the hands of the ultimate consumer (Christopher, 1992).
- Networks of manufacturing and distribution sites that procure raw materials transform them into intermediate and finished products and distribute the finished products to customers (Lee & Billington, 1992).
- The set of entities, including suppliers, logistics services providers, manufacturers, distributors, and resellers flow through materials, products, and information (Kopczak, 1997).
- A network of entities starts with the suppliers' supplier and ends with the customers' customizing the production and delivery of goods and services (Lee & Ng, 1997).

Another aspect of SCM is communication. In fact, the National Initiative for Supply Chain Integration was created to improve and standardize communication and business processes through manufacturing supply chains and to share the results. Moreover, supply chain management and logistics are often viewed as overlapping. For example, the Council of Logistics Management (CLM) defines logistics as a process of planning, implementing, and controlling the efficient, adequate flow and storage of goods, services, and related information from the point of origin to the point of consumption to conform to customer requirements (Lummus et al., 2001). The distribution and logistics process determines how products are retrieved and transported from the storage warehouse to retailers. The process also includes the management of inventory retrieval, transportation, and final product delivery. The interaction of these processes produces an integrated supply chain (Haghani & Afshar, 2009).

A key point in supply chain management is that the entire process must be viewed as one system. Integrated supply chain management is about going from the external customer and then managing all the processes needed to provide the customer with value horizontally (Li et al., 2006). In fact, Jahre and Jensen (2010) expound on horizontal and vertical logistics coordination. The horizontal coordination involves actors, activities, and resources at strategic, tactical, and operational levels in information, money, and material flows for companies at the "same" stage in the supply chain. The focus is on the companies and their specific tasks. The vertical coordination revolves around actors, activities, and resources at strategic, tactical, and operational levels in companies' information, money, and material flow at the "different" stages in the supply chain. The focus is on the customer and synchronization.

Moreover, horizontal coordination is done for economic scaling and to reduce costs for the individual company. It enables access to more physical resources, information, and know-how competence. Vertical coordination is for the reduction of overall supply chain costs but can increase costs for some actors. It tends to improve customer service by creating easier flows.

Because of the intricate relationships within a supply chain network, each entity's performance is dependent on the performance of other entities as well as the ability (and willingness) to coordinate activities within the supply chain. An essential component in the design and analysis of the supply chain is the establishment of appropriate key performance measures (Stumpe & Katina, 2019) that determine the efficiency and effectiveness of the existing system. These measures can also be used to make a comparison to the competing alternative systems. The performance measures can also be used to design proposed systems by determining the values of the decision variables that yield the most desirable performance levels. There are a variety of performance measures. Beamon (1998) suggests qualitative, quantitative, and customer-based measures. Qualitative performance measures elements of flexibility, information and material flow integration, effective risk management, and supplier performance. Quantitative performance measures are based on cost and profit objectives—these tend to include elements of cost minimization, sales maximization, profit maximization, inventory minimization, and return on investment maximization. The third measure is a customer-based responsiveness objective and includes such elements as fill-rate maximization, product lateness minimization, customer response time minimization, lead time minimization, and function duplication minimization (Beamon, 1998).

Sources of risks are relevant to supply chains. These can be organized into three categories: external to supply chain, internal to supply chain, and network-related (Brindley, 2017). Arguably, modern supply chains are vulnerable to various risks due to complex networks stretched over multiple geographical locations (Aqlan & Lam, 2015; Keating et al., 2022). And while the core factors of supply chain profitability revolve around responsiveness, efficiency, and reliability, there is a need for resilience in supply chains. In this case, supply chain resilience is seen as the property of a supply chain network that enables it to regain its original configuration soon after disaster disruption (Papadopoulos et al., 2017).

Interestingly, according to the Business Continuity Institute, a network of disaster recovery and business continuity experts, in 2017, 65% of companies experienced at least one supply chain disruption, leading to a loss of productivity, decrease in customer service, and loss of revenue. Moreover, during the past two decades, supply chain disruptions caused by natural and artificial disasters occurred more frequently, with greater intensity, and had severe consequences (Ivanov et al., 2016). The disruptive risks are distinguished by rare occurrences and high-performance impact, which disturb network structure and critical performance metrics. A vital phenomenon connected to disruption is the ripple effect. The ripple effect is a testimony to the intricate interdependencies within the supply chain. Six categories of interdependency are suggested (Dudenhoeffer et al., 2006; Katina et al., 2014; Rinaldi et al., 2001):

- *Physical Interdependency:* This is a relationship that "arises from the physical linkage between the inputs and outputs of two agents [where the] commodity produced or modified by one infrastructure (an output) is required by another infrastructure for it to operate (an input)" (Rinaldi et al., 2001, p. 15) such as drinking water and electricity.
- *Cyber Interdependency:* A relationship based on the ubiquitous and pervasive use of information and communications technologies (ICT). Supply chain systems provide essential goods and services with the help of control systems such as supervisory control and data acquisition (SCADA) systems that remotely monitor and control operations using coded signals over ICT communication channels (Rinaldi et al., 2001).
- *Geographical Interdependency:* This is a relationship that exists when different supply chain systems share the same environment, such as electrical power lines that share the same corridor with a bridge (DiSera & Brooks, 2009).
- *Logical Interdependency:* A logical interdependency exists in a supply chain if the state of one system depends on the state of the other via some mechanism that is neither physical, cyber, nor geographical (Rinaldi, 2004) such as power deregulation policy.
- *Policy and/or Procedural Interdependency:* This is a "hidden" and not-so-obvious relationship that only becomes apparent after a change, in the form of a policy and/or procedure that takes effect in one part of the system. For example, several regulations that were issued in the wake of the 9/11 attacks affected all air transport systems, changing the flying experience (Mendonca & Wallace, 2006).
- *Societal Interdependency:* The situation in which public opinion affects supply chain operations. For example, after the 9/11 attacks, air traffic was reduced due to the public's evaluation of travel safety, resulting in job cuts and bankruptcies (Dudenhoeffer et al., 2006).

Under these circumstances, the absence of capacity or inventory at the disrupted facility may lead to a lack of material in the affected area. Hence, the notion of resilience in supply chain networks is necessary since it allows the network to prepare for and

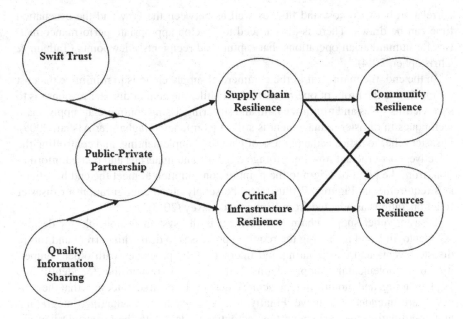

FIGURE 2.6   Resilience in supply chain management.

adapt to unexpected events. Figure 2.6 is adapted from Papadopoulos et al. (2017) and developed to depict how resilience can be added to a supply chain network.

2.4   DISASTER SUPPLY CHAIN

At the heart of any disaster relief system is the disaster, an event that forms the focal point around which all the supply chain management activities are organized. In response to a specific event, momentary supply chains are finite-life chains that are often deployed. Accordingly, all life cycle stages (of a supply chain) need to be considered in a relatively short period (Day et al., 2012). However, embracing supply chain management as a core function of disaster response is relatively new (Tomasini & Van Wassenhove, 2009). And yet the difference between supply chain management and disaster relief operation is the involvement of the intermediate channels (Kumar & Havey, 2013). Furthermore, several studies emphasized that some supply chain concepts share similarities to emergency logistics, suggesting that methods and tools developed for commercial supply chains can be successfully adapted to emergency response logistics despite the strategic goals of the commercial supply chain being different from disaster response logistics. For example, Beamon (2004) and Thomas and Kopczak (2005) articulate supply chain management capabilities within humanitarian organizations. Altay et al. (2009) suggest a need to improve communications (internal and external), invest in long-term relationships, develop cross-functional teams, and promote trust and commitment in the logistics chain. Nahum et al. (2017)'s research suggests that evacuation route issues in both cases can involve aspects of multi-objective and stochasticity, and as such, a positive correlation between cost and flow as well as between the flow and the evacuation time can be drawn. There is also a need to develop appropriate performance metrics for humanitarian operations that capture aid recipients' viewpoints (Tatham & Christopher, 2014).

In the end, the main goal in the commercial supply chain is to minimize the cost (or maximize the profit of operations). Meanwhile, the goal of disaster response is to save victims' lives and minimize pain and suffering. Using commercial supply chain techniques in disaster management is still in its infancy (Haghani & Afshar, 2009). Disaster relief logistics is the process of planning, implementing, and controlling the effective, cost-efficient flow and storage of goods and materials and related information, from the point of origin to the point of consumption to meet the end beneficiary's requirements. Figure 2.7 illustrates the supply chain configuration for disaster relief operations as adapted from Kumar and Havey (2013).

Disaster relief supply chain management is the system responsible for designing, deploying, and managing the required processes to deal with current and future disaster events and coordinating and interacting the processes with other competitive or complementary supply chains. Moreover, it is responsible for identifying, implementing, and monitoring the achievement of the desired outcomes that the processes are intended to achieve. Finally, it is responsible for evaluating, integrating, and coordinating the various parties' activities to deal with the events (Day et al., 2012). Disaster relief logistics is more tactical, operational, and execution-oriented compared to disaster relief supply chains. Day et al. (2012) research suggests that all related supply chain management, disaster relief logistics, and disaster relief supply chain management are related (Figure 2.8). Disaster supply chain management is characterized by large-scale operations, unusual constraints, irregular demand, and unreliable or nonexistent supply and transportation information. The engineering of a distribution network is challenging because of the nature of the unknown (locations, type, spread, and magnitude of events, politics, and culture). Disaster management organizations must deal with zero lead time in their supply chain, and agile capabilities are essential in this aspect. Meanwhile, collaboration is crucial

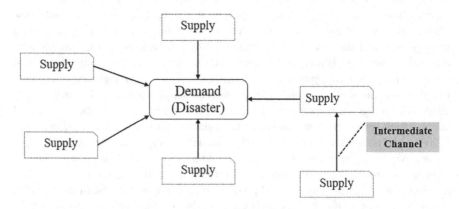

FIGURE 2.7   A configuration of the supply chain for disaster relief operation.

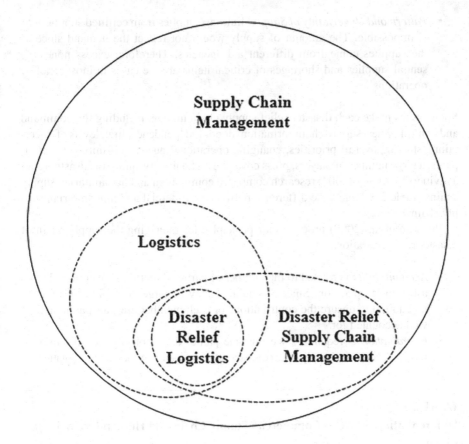

FIGURE 2.8   Disaster relief in the context of supply chain management.

in disaster management since no single entity has sufficient resources to respond (Scholten et al., 2014).

Therefore, one can consider a disaster relief operation supply chain to be like a commercial supply chain that needs effective delivery, involving a complex supply network, uncertainty, and visibility. Other differences include the following (Pujawan et al., 2009):

- *Configuration of the Supply Network:* disaster relief supply chain has only one demand point and is achieved from multiple sources of supply. Configuration is unstable, and most suppliers only supply once during the relief operation.
- *The Number of Involved Parties:* in case of a disaster, most relationships are developed rapidly. And since many parties are volunteers, their participation may be without any evaluation. And as such, each party contributes uniquely.
- *Pattern and Uncertainty of Demand:* the demand is mainly unpredictable, short-term in nature, and volatile with time, and minimal past data can be used to evaluate the level of demand.

- *Pattern and Uncertainty of Supply:* limiting supplies from certified suppliers is impossible. The amount of supply would not reflect the demand since the supplies come from different aid agencies. Therefore, excess nonessential supplies and shortages of critical items are the cases in most relief operations.

Some issues make each disaster relief supply chain unique, including the command and control issues, supply chain formation, donor independence, high levels of uncertainty, shifting overall priorities, changing operational needs, self-initiated participants, a large number of players, press coverage and publicity, and post-disaster relief activities. Beamon's (2004) research compared commercial and humanitarian supply chains. Table 2.3 depicts the difference between commercial and humanitarian supply chains.

Pujawan et al. (2009) propose four principles for managing the supply chain of disaster relief operations:

- *Accountability* is related to the number of parties directly involved in a disaster relief operation. Since the number may be large, a significant effort is required to ensure the contributions of each member are transparent and well accounted for.
- *Coordination* is required since multiple parties are involved, and objectives may conflict in the supply chain disaster relief operation. A lack of coordination

---

TABLE 2.3

Differentiating between Commercial Supply Chain and Humanitarian Supply Chain

Distinctive	Commercial Supply Chain	Humanitarian Supply Chain
Distribution network configuration	Well-defined methods for determining the number and locations of distribution centers	Challenging due to the nature of the unknowns (e.g., locations, type, and size of events, politics, and culture)
Information system	Generally well-defined, using advanced technology	Information is often unreliable, incomplete, or non-existent
Inventory control	Utilizes well-defined methods for determining inventory levels based on the lead time, demand, and target customer service levels	Inventory control is challenging due to the high variations in lead times, requirements, and demand locations
Lead time	Lead time is determined by the supplier-manufacturer-DC-retailer	Lead time between the occurrence of the demand and the need for the demand is zero
Strategic goals	Typically to produce high-quality products at low cost to maximize profitability	Minimize loss of life and alleviate suffering
What is "demand"	Products	Emergency supplies, equipment, and personnel

can lead to excess supply for some types of goods, shortage of other types, and duplication of efforts. The humanitarian agencies may not be aware of their counterparts' activities, which may lead to duplication and low-resource utilization in many supply chain decisions, including warehousing, transportation, and procurement.

- *Information visibility* is related to visibility in disaster relief operations. Critical information must be well accessible by the interested parties. In case of a disaster, there should be information about the available items, the on-hand quantity, the expected supply days, and the location. In order to have this information, there is a need for accurate recording of inventory transactions, and regular reporting to the public, aid agencies, and major donors are required.
- *Professionalism* is related to the availability of well-trained people to perform the tasks and the standard operating procedure to follow. Promoting professionalism in the disaster supply chain and logistics management is complex because the turnover rate of people working on relief projects is high. Moreover, professionalism involves standard operating procedures being compiled during the relief operation.

All the involved parties should maintain the principles of equality and impartiality in delivering aid to the affected areas. Also, the previously mentioned principles do not operate independently of one another. For example, the visibility of information can help in creating better supply chain coordination.

The emergency supply chain objectives for disaster management include, among others, optimal deployment of military units, resources and equipment, water supply, food, clothing, infrastructure reconstruction, and medical support (Ben Othman et al., 2017). However, cost minimizing, delivery delays, and environment complexity (i.e., number of actors and uncertainty) are relevant factors to consider in developing automatic models and tools to optimize logistics solutions to meet emergency needs (Ben Othman et al., 2017). Such models and tools can be instrumental in decision-making, especially for those who must make real-time decisions. However, the main problem in disaster relief operations is in the distribution of the items quickly and in sufficient quantity for the affected areas (Thomas & Kopczak, 2005). The blockage in distributing the supplies might be the damaged infrastructure and unavailability of accurate information about the required amounts. The oversupply of nonessential goods could slow the logistical response. And yet another critical challenge in the disaster relief supply chain is the flow of resources (Day et al., 2012).

Moreover, there is a tendency to converge and offer assistance to individuals (or groups) at (or near) the site of the disaster, especially to those that are physically (or emotionally) unable to take care of themselves. However, convergence can expose the supply chain to additional unique challenges. For example, convergence tends to incorporate groups not traditionally involved in disaster response or mitigation decision-making, which could exacerbate rescue and recovery efforts (Day et al., 2012). It can also affect boundaries, change the scope of performance standards, and hinder trust within the supply chain (Day et al., 2012).

Critical areas of the disaster relief supply chain identified by experienced practitioners are categorized in four themes: (i) demand signal visibility and requirements determination, (ii) information management and relief activity coordination, (iii) disaster relief planning, and (iv) managing relationships and developing trust along the supply chain (Day et al., 2012).

Moreover, Perry (2007) suggests that local knowledge is helpful in assessing the needs considering culture and traditions. Altay and Green (2006) review the literature to identify potential directions for researching disaster operations and provide a database of practical resources. After a disaster, resilience in the supply chain determines the path to normalcy through various collaborations in the supply chain networks. It is also understood that there is a need for developing frameworks for the disaster supply chain. For example, Ransikarbum (2015) argues that the existing models for post-disaster disruption management are limited, and an integrated system view is required. The model for multi-objective integrated response and recovery (MOIRR) proposed by Ransikarbum and Mason (2016) includes three objectives that integrate the supply disruption problem that occurs during a disaster: First, provide a model to maximize equity or fairness. Second, minimize total unsatisfied demand across the demand/beneficiary nodes. Third, minimize the total network cost as the fund is used to restore disrupted nodes, restore disrupted arcs, and transport supply units based on origin-destination pair information. Alderson et al. (2014) illustrate how to build and solve a sequence of models to assess and improve the resilience of an infrastructure system after disruptive events. Akgün et al. (2015) developed a pre-disaster phase model to locate prepositioned supplies close to areas prone to disaster. This approach enables the creation of a relatively reliable facility network to minimize response time in times of need. Cui et al. (2011) develop a goal programming model for a web service problem and suggest that including real constraints only is insufficient for describing user requirements; further, there may be no way to satisfy all real constraints simultaneously (Ransikarbum, 2015). Jahre and Jensen (2010) have proposed applying the cluster concept to humanitarian logistics coordination. Some platforms assist in case of disasters. For example, Humanitarian Aid Distribution System is a web-based decision support platform designed to aid non-experienced users in making more effective decisions during the response phase (Vitoriano et al., 2009, Ortuño et al., 2011). However, Ben Othman et al. (2017) contend that most literature used centralized planning approaches for emergency supply chain, while a distributed planning approach can be more suitable according to the large and distributed nature of the emergency supply chain.

2.4.1 Disaster Supply Communication and Trust

Trust is a crucial challenge influencing collaboration and communication in disaster situations. Trust plays a vital role in supply chain relationships, and research suggests that improving trust is possible (Murayama et al., 2013). Two dimensions can be used to define trust: affective and cognitive. Affective trust is loyalty to a partner. It is typically developed through firsthand experiences. Cognitive trust reflects a choice rationale and is often based on the economic considerations of the partners participating in the supply chain (Day et al., 2012). While it is imperative to develop trust in the disaster relief supply chain, a systematic approach is also necessary to aid in the prioritization of needs. And once all participants are aware of this prioritization, it makes it easier for them to provide the necessary resources. Swift trust has a crucial role in coordination improvement among humanitarian actors. Tatham and Kovács (2010) suggest that swift characteristics of trust are (i) information regarding actors involved in disaster relief activities, (ii) dispositional trust, (iii) clear rules for classification of processes and procedures, (iv) role clarity, and (v) use of category (e.g., gender, ethnicity).

It is widely known that most of the research in emergency disaster management focuses on disaster results, sociological impacts on communities, psychological effects on survivors or rescue teams, and organizational design and communication problems (Altay & Green, 2006). Also, operations research and management science (OR/MS) are popular approaches in the previous research. However, there is a scarcity of research suggesting how to integrate interrelated problems of a large-scale multimodal network flows (Alipour, 2021). Moreover, Nahum et al. (2017) argue that most emergency response operations research is focused on evacuation problems from the perspective of transportation modelings, such as network design and traffic assignment. Bai (2016) argues that the current literature regarding emergency supplies mainly depends on deterministic optimization methods. The issue with deterministic optimization methods is that there are cases where fixed distributions cannot be determined in the problem, and adopting the obtained solution is unsuitable for emergency management. One of the significant problems faced immediately after a disaster is transporting large amounts of multiple commodities, including food, clothing, medicine, medical supplies, machinery, and personnel, from several origins to several destinations. In such cases, transportation must be quick and efficient to maximize the survival rate of the affected population and minimize the cost of such operations.

2.4.2 EMERGENCY LOGISTICS

Ben Othman et al. (2017) suggest that emergency logistics is a set of logistics actors that interact and coordinate to accomplish emergency logistics requirements with the main features, including the following:

- Automated emergency logistics systems for efficient disaster relief supply and recovery
- Supply resources and workforce assessment to adjust to unexpected difficult circumstances
- Decentralized system to deal with the uncertainties

Again, a disaster area is a dynamic environment that might evolve over time. Therefore, emergency logistics have to be seen as time-sensitive operations. The lack of information about the available infrastructure, supplies, and requirements would complicate this dynamic environment even more. Victims' high stake of life or death requires high accuracy and tractability. Logistics planning in emergencies includes sending several relief supplies from different sources to multiple distribution points within the disaster region.

Moreover, relief supplies are sent and distributed within the supply chain and logistics structure. This structure and the accompanying decisions—the right type of relief, number of supplies, routes, and transportation mode—are essential in minimizing the suffering of the disaster victims. Some of the demand items are one-time demand, and some are subject to expiration. The demand information might not be completed and accurate initially (Haghani & Afshar, 2009). Logistic coordination in disasters includes selecting sites that can maximize the coverage of affected areas and minimize the delays in supply delivery operations. Again, we see that coordination and cooperation are essential, especially between transportation modes that are required for managing the response operations.

2.4.3 HUMANITARIAN LOGISTICS

Humanitarian logistics is a process of planning, managing, and controlling the flow of information, relief, and services. The flow is typically from the point of origin to the points of urgent needs, usually the location under emergency conditions (Ransikarbum, 2015). Based on the different types of technologies described, there are three activity structures or value creation models: chain, shop, and network. Network taxonomies are based on some aspects, including who takes part, controls, the type of coordination, the purpose, the products, and the degree of dynamics, among others. This type of network is temporary, unique, and created only for humanitarian operations. The network should be tailored based on the uncertainties about the time, needs, and available infrastructures (Jahre et al., 2009). Based on a literature review of humanitarian logistics research, Ransikarbum (2015) argues that more research is required in humanitarian logistics. The performance of humanitarian logistics is based not only on the effectiveness but the efficiency of post-disasters activities as well. Humanitarian logistics can be critical in devising improved ways of managing multi-stakeholder relief operations (Ransikarbum, 2015). Moreover, humanitarian logistics vary from commercial logistics. For example, while humanitarian and commercial logistics processes account for the flows of goods between the nodes of a network, commercial supply chains tend to be a permanent fixture in society. Humanitarian logistics tend to be temporary and only set up after a disaster (Gösling & Geldermann, 2014).

Tomasini and Van Wassenhove (2009) suggest that there are three stages—ramp-up, sustain, and ramp-down—in disaster response. The first few days after a disaster strike is covered in the ramp-up stage, where speed is the main driver, and lead time reduction is also essential. The focus is on accessing the field and setting up operations according to the highest priority. The agencies' objective is to get to the affected area, observe and document the damage extent, assess needs, and implement urgent solutions. In the sustain stage, agencies try to implement programs where cost and efficiencies are more concerned.

During the ramp-down stage, the operations are transferred to the local actors. In effect, this is the "exit strategy." Once the operation has been set up, roles have been defined, and visibility of the process to assist the beneficiaries has increased, the cost would be the adopted driver. The cost and speed balance are also important in disaster preparedness, where agencies try to develop better-prepared processes and products to achieve both aims of cost and speed. In the case of a combination of specialized and independent organizations, problems arise related to coordination, such as the 2005 Indian Ocean tsunami and the 2004–2005 Darfur crisis. There were overlapping in some provisions of relief, some populations were not appropriately considered, and prioritization of pipelines was very problematic. These facts indicate the coordination requirement in contingency planning, needs assessment, appeals, and transport management. The concern would be observing when and how the key players should collaborate and coordinate (Jahre & Jensen, 2010).

Aid disaster relief (i.e., funds, goods, human resources, knowledge, and expertise), which is the source of relief, will tend to come from private organizations. In recent times, there is also a trend for private companies to design their social engagement via a long-term partnership with humanitarian organizations for charitable concerns and as an opportunity for learning and developing their businesses. Alternatively, humanitarian agencies invest equal resources to enhance their performance and core competencies in interacting with the private sector. The main benefit is back-office support for better preparation and key asset movement during a disaster (Papadopoulos et al., 2017). Balcik et al. (2010) describe potential relief chain coordination mechanisms' attributes, costs, and applicability (Table 2.4). These mechanisms are derived from the commercial supply chain coordination mechanisms.

Still, the investigation needs in humanitarian supply chain and relief operations include (i) distribution planning, (ii) information and communication system, (iii) sourcing and supplier management, (iv) performance measurement, and (v) transportation, mode choice, and routing (Ransikarbum, 2015). Additionally, decision-making in humanitarian logistics operations is dynamic. Therefore, applying dynamics simulations to study the interactions and effects of stakeholders is warranted (Ransikarbum, 2015). To have more efficient coordination and integration within a supply chain and among multiple supply chains, there is a need to use a decentralized structure for the distribution of activities, and resources are an attractive part of the preparedness strategy for many multinational NGOs. The decentralized units are represented as information intermediaries between players to be activated in the response phase by instituting business relations (Jahre et al., 2009).

In the end, supply chain management (SCM) handles activities between separate entities, and logistics focuses on the internal movement of goods. SCM supports all purchasing, production, and distribution of goods. Logistics, meanwhile, moves and stores goods between different points in the supply chain. To learn more, read our article on inbound and outbound logistics. Logistics focuses on the movement and storage of items in the supply chain. SCM is more comprehensive, covering all of the coordination between partners that have a role in this network, including sourcing, manufacturing, transporting, storing, and selling. The ultimate goal of SCM is to find processes that ensure a smooth, efficient flow of goods that give customers an excellent experience and drive the business forward.

Moreover, the supply chain logistical components can be used to manage goods or services. Each element helps move materials, finished goods, and services via the many steps in the supply chain. The following are the logistical components of supply chains (Jenkins, 2021):

TABLE 2.4

Evaluating Potential Relief Chain Coordination Mechanisms in Supply Chains

Relief Coordination Mechanism	Currently Observed	Coordination Costs	Opportunistic Risk Cost	Operational Risk Cost	Technological Requirements for NGO	Conductive to Relief Environment	Implementation Potential
Collaborative	Yes	Low	Low	Low	Low	Yes	NGOs but low overall High
Procurement warehouse	No	High	Varies	High	Medium	Yes	Low
QR, CR, VMI, CVMI	No	Low	High	High	High	No	Higher for large
Shipper collaboration 4PL	No	High	High	High	Medium	No	Low
Standardization third-party	Yes	Low	Low	Low	Low	Yes	High
Third-party Warehousing (umbrella org)	Yes	Medium	Medium	Varies	Low	Yes	High
Warehousing (private sector partner) Transportation	No	High	Varies	High	Medium	No	Low

- *Information:* Information helps track the status of items and all supply chain processes, informing business decisions at each step.
- *Storage:* Storage is the practice of holding supplies in the correct quantity and suitable location. Businesses must balance demand and supply to prevent overstock and out-of-stock situations.
- *Warehousing:* This component controls the day-to-day warehouse operations, such as receiving, put-away, picking, packing, shipping, and receiving.
- *Material Handling:* Material handling can refer to the limited movement of items within a building or a delivery vehicle. Others extend the definition to include the storage, security, and transfer of goods throughout the manufacturing, distribution, and delivery processes.
- *Packaging:* Proper packaging ensures items arrive undamaged and ship for the lowest possible cost.
- *Unitization:* Unitization makes items efficient to arrange, transport, and store. Unitization methods also ensure that material handling equipment can move items efficiently without damaging them. The cube is one of the most accessible units to store and shift, so it's a popular type of unitization.
- *Inventory Control:* Inventory control incorporates storage and warehousing techniques to optimize the types and amount of stock held and where. Companies can use inventory management formulas to calculate demand better.
- *Transportation:* This component is responsible for moving goods along the supply chain to the next node or directly to the customer. Transportation modes include cars, trains, trucks, planes, and ships.

And according to Abby Jenkins of Oracle Netsuite, if one takes the view that SCM only addresses planning, sourcing, manufacturing, delivery, and returning, then the role of logistics within supply chain management can also be summarized (Jenkins, 2021):

- Planning = information inventory control
- Sourcing = information material handling transportation
- Manufacturing = information material handling storage warehousing packaging/unitization transportation
- Delivering = information warehousing inventory management materials-handling packaging transportation
- Returning = information packaging/unitization material handling transportation warehousing

CONCLUSIONS

It is no secret that disasters are occurring more frequently, becoming less deadly and therefore more costly. The number of catastrophes and people affected by disasters is rising. In fact, it appears that the frequency and magnitude of failures and the subsequent impacts will only increase (Rasmussen and Batstone, 1989).

At the same time, there is evidence that developing nations are (and will continue to be) excessively impacted by disaster outcomes. Under the traditional approach, communications and decision-making are centralized, yet supply disaster challenges can be broken down into such issues as dimensions, understanding, and customs. In these scenarios, using a central command and control approach (within a dynamic situation) will only lead to more problems. In fact, it can create restrictions. In turn, such restriction creates an environment in which developing an enhanced decision-making process is impossible. In a dynamic disaster situation, a full range of information (and knowledge) is never available. In such conditions, a distributed decision-making approach will be more efficient within complex and dynamic organizations. There remains a need for flexible and innovative models and tools supporting those who must make critical decisions, especially in disaster real-time situations.

REFERENCES

Akbari, M. E., Farshad, A. A., & Asadi-Lari, M. (2004). The devastation of Bam: An overview of health issues 1 month after the earthquake. *Public Health, 118*(6), 403–408. https://doi.org/10.1016/j.puhe.2004.05.010

Akgün, İ., Gümüşbuğa, F., & Tansel, B. (2015). Risk based facility location by using fault tree analysis in disaster management. *Omega, 52*, 168–179. https://doi.org/10.1016/j.omega.2014.04.003

Alderson, D. L., Brown, G. G., & Carlyle, W. M. (2014). Assessing and improving operational resilience of critical infrastructures and other systems. In A. M. Newman, J. Leung, J. C. Smith, & H. J. Greenberg (Eds.), *Bridging data and decisions* (pp. 180–215). INFORMS. https://doi.org/10.1287/educ.2014.0131

AliPour, F. S. (2021). *Application of a blockchain enabled model in disaster aids supply network resilience* [Dissertation, Old Dominion University Libraries]. https://doi.org/10.25777/FKR7-A212

Altay, N., & Green, W. G. (2006). OR/MS research in disaster operations management. *European Journal of Operational Research, 175*(1), 475–493. https://doi.org/10.1016/j.ejor.2005.05.016

Altay, N., Prasad, S., & Sounderpandian, J. (2009). Strategic planning for disaster relief logistics: Lessons from supply chain management. *International Journal of Services Sciences, 2*(2), 142–161. https://doi.org/10.1504/IJSSci.2009.024937

Apte, S., & Petrovsky, N. (2016). Will blockchain technology revolutionize excipient supply chain management? *Journal of Excipients and Food Chemicals, 7*(3), 76–78.

Aqlan, F., & Lam, S. S. (2015). A fuzzy-based integrated framework for supply chain risk assessment. *International Journal of Production Economics, 161*, 54–63. https://doi.org/10.1016/j.ijpe.2014.11.013

Bai, X. (2016). Optimal decisions for prepositioning emergency supplies problem with Type-2 fuzzy variables. *Discrete Dynamics in Nature and Society, 2016*, e9275192. https://doi.org/10.1155/2016/9275192

Balcik, B., & Beamon, B. M. (2008). Facility location in humanitarian relief. *International Journal of Logistics Research and Applications, 11*(2), 101–121. https://doi.org/10.1080/13675560701561789

Balcik, B., Beamon, B. M., Krejci, C. C., Muramatsu, K. M., & Ramirez, M. (2010). Coordination in humanitarian relief chains: Practices, challenges and opportunities. *International Journal of Production Economics, 126*(1), 22–34. https://doi.org/10.1016/j.ijpe.2009.09.008

Beamon, B. M. (1998). Supply chain design and analysis: Models and methods. *International Journal of Production Economics*, 55(3), 281–294. https://doi.org/10.1016/S0925-5273(98)00079-6

Beamon, B. M. (2004). Humanitarian relief chains: Issues and challenges. Proceedings of the 34th International Conference on Computers and Industrial Engineering. http://courses.washington.edu/ie59x/abstracts/IEseminar05.pdf

Ben Othman, S., Zgaya, H., Dotoli, M., & Hammadi, S. (2017). An agent-based decision support system for resources' scheduling in emergency supply chains. *Control Engineering Practice*, 59, 27–43. https://doi.org/10.1016/j.conengprac.2016.11.014

Berry, D., Towill, D. R., & Wadsley, N. (1994). Supply chain management in the electronics products industry. *International Journal of Physical Distribution & Logistics Management*, 24(10), 20–32. https://doi.org/10.1108/09600039410074773

Birkmann, J., & von Teichman, K. (2010). Integrating disaster risk reduction and climate change adaptation: Key challenges—scales, knowledge, and norms. *Sustainability Science*, 5(2), 171–184. https://doi.org/10.1007/s11625-010-0108-y

Brindley, C. (Ed.). (2017). *Supply chain risk*. Routledge. https://doi.org/10.4324/9781315242057

Cartier, A. V., Laprade, C. A., Pierri, M. V., & Worsham, D. H. (2009). *Disaster decision making: Hurricanes Katrina and Gustav in New Orleans* [Project Number: 0902, Worcester Polytechnic Institute]. https://web.wpi.edu/Pubs/E-project/Available/E-project-031609-172502/unrestricted/IQPReportFINAL.pdf

Casti, J. (2012). *X-Events: Complexity overload and the collapse of everything*. William Morrow.

Cavdur, F., Kose-Kucuk, M., & Sebatli, A. (2016). Allocation of temporary disaster response facilities under demand uncertainty: An earthquake case study. *International Journal of Disaster Risk Reduction*, 19, 159–166. https://doi.org/10.1016/j.ijdrr.2016.08.009

CCAHA. (2020). *Introduction to the incident command system*. Conservation Center for Art & Historic Artifacts. https://ccaha.org/resources/introduction-incident-command-system

Chang, S. E. (2010). Urban disaster recovery: A measurement framework and its application to the 1995 Kobe earthquake. *Disasters*, 34(2), 303–327. https://doi.org/10.1111/j.1467-7717.2009.01130.x

Christopher, M. (1992). *Logistics and supply chain management*. Irwin Professional Publishing.

Chylek, L. A., Harris, L. A., Tung, C.-S., Faeder, J. R., Lopez, C. F., & Hlavacek, W. S. (2014). Rule-based modeling: A computational approach for studying biomolecular site dynamics in cell signaling systems. *Wiley Interdisciplinary Reviews. Systems Biology and Medicine*, 6(1), 13–36. https://doi.org/10.1002/wsbm.1245

Colicchia, C., & Strozzi, F. (2012). Supply chain risk management: A new methodology for a systematic literature review. *Supply Chain Management: An International Journal*, 17(4), 403–418. https://doi.org/10.1108/13598541211246558

Coppola, D. P. (2006). *Introduction to international disaster management*. Elsevier.

Croom, S., Romano, P., & Giannakis, M. (2000). Supply chain management: An analytical framework for critical literature review. *European Journal of Purchasing & Supply Management*, 6(1), 67–83. https://doi.org/10.1016/S0969-7012(99)00030-1

Cui, L., Kumara, S., & Lee, D. (2011). Scenario analysis of web service composition based on multi-criteria mathematical goal programming. *Service Science*, 3(4), 280–303. https://doi.org/10.1287/serv.3.4.280

Day, J. M., Melnyk, S. A., Larson, P. D., Davis, E. W., & Whybark, D. C. (2012). Humanitarian and disaster relief supply chains: A matter of life and death. *Journal of Supply Chain Management*, 48(2), 21–36. https://doi.org/10.1111/j.1745-493X.2012.03267.x

DiSera, D., & Brooks, T. (2009). The geospatial dimensions of critical infrastructure and emergency response. *Pipeline and Gas Journal*, 236(9), 1–4.

Dudenhoeffer, D., Permann, M. R., & Manic, M. (2006). CIMS: A framework for infrastructure interdependency modeling and analysis. *Proceedings of the 38th Conference on Winter Simulation*, 478–485. https://doi.org/10.1109/WSC.2006.323119

Ellram, L. M. (1991). Supply-chain management: The industrial organisation perspective. *International Journal of Physical Distribution & Logistics Management, 21*(1), 13–22. https://doi.org/10.1108/09600039110137082

FEMA. (1996). *Glossary of terms: Guide for all-hazard emergency operations planning* (GLO 101). Federal Emergency Management Agency. www.fema.gov/pdf/plan/glo.pdf

FEMA. (2001). *Robert T. Stafford Disaster Relief and Emergency Assistance Act, PL 100–707, signed into law November 23, 1988; amended the Disaster Relief Act of 1974, PL 93–288, no. PL 100–707, 2001.* www.fema.gov/disaster/stafford-act

Fogli, D., Greppi, C., & Guida, G. (2017). Design patterns for emergency management: An exercise in reflective practice. *Information & Management, 54*(7), 971–986. https://doi.org/10.1016/j.im.2017.02.002

Garcia, D. J., & You, F. (2015). Supply chain design and optimization: Challenges and opportunities. *Computers & Chemical Engineering, 81*, 153–170. https://doi.org/10.1016/j.compchemeng.2015.03.015

Gerber, B. J. (2007). Disaster management in the United States: Examining key political and policy challenges. *Policy Studies Journal, 35*(2), 227–238. https://doi.org/10.1111/j.1541-0072.2007.00217.x

Glik, D. C. (2007). Risk communication for public health emergencies. *Annual Review of Public Health, 28*, 33–54. https://doi.org/10.1146/annurev.publhealth.28.021406.144123

Gösling, H., & Geldermann, J. (2014). A framework to compare OR models for humanitarian logistics. *Procedia Engineering, 78*, 22–28. https://doi.org/10.1016/j.proeng.2014.07.034

Gunes, A. E., & Kovel, J. P. (2000). Using GIS in emergency management operations. *Journal of Urban Planning and Development, 126*(3), 136–149. https://doi.org/10.1061/(ASCE)0733-9488(2000)126:3(136)

Haddow, G., Bullock, J., & Coppola, D. P. (2017). *Introduction to emergency management.* Butterworth-Heinemann. www.elsevier.com/books/introduction-to-emergency-management/bullock/978-0-12-817139-4

Haghani, A., & Afshar, A. M. (2009). *Supply chain management in disaster response* (Grant DTRT07-G-0003). Mid-Atlantic Universities Transportation Center. https://rosap.ntl.bts.gov/view/dot/34430

Hale, T., & Moberg, C. R. (2005). Improving supply chain disaster preparedness: A decision process for secure site location. *International Journal of Physical Distribution & Logistics Management, 35*(3), 195–207. https://doi.org/10.1108/09600030510594576

Harland, C. M. (1996). Supply chain management: Relationships, chains and neetworks. *British Journal of Management, 7*(S1), S63—S80. https://doi.org/10.1111/j.1467-8551.1996.tb00148.x

Hill, M. (2014). *Studying public policy: An international approach.* Policy Press.

Holguín-Veras, J., Pérez, N., Ukkusuri, S., Wachtendorf, T., & Brown, B. (2007). Emergency logistics issues affecting the response to Katrina: A synthesis and preliminary suggestions for improvement. *Transportation Research Record, 2022*(1), 76–82. https://doi.org/10.3141/2022-09

Hu, H., Lei, T., Hu, J., Zhang, S., & Kavan, P. (2018). Disaster-mitigating and general innovative responses to climate disasters: Evidence from modern and historical China. *International Journal of Disaster Risk Reduction, 28*, 664–673. https://doi.org/10.1016/j.ijdrr.2018.01.022

Imran, M., Castillo, C., Diaz, F., & Vieweg, S. (2015). Processing social media messages in mass emergency: A survey. *ACM Computing Surveys, 47*(4), 67:1–67:38. https://doi.org/10.1145/2771588

Ivanov, D., Tsipoulanidis, A., & Schönberger, J. (2016). *Global supply chain and operations management: A decision-oriented introduction to the creation of value.* Springer.

Jahre, M., & Jensen, L. (2010). Coordination in humanitarian logistics through clusters. *International Journal of Physical Distribution & Logistics Management, 40*(8/9), 657–674. https://doi.org/10.1108/09600031011079319

Jahre, M., Jensen, L., & Listou, T. (2009). Theory development in humanitarian logistics: A framework and three cases. *Management Research News*, *32*(11), 1008–1023. https://doi.org/10.1108/01409170910998255

Jenkins, A. (2021). Supply chain vs logistics: What's the difference? *Oracle NetSuite*. www.netsuite.com/portal/resource/articles/erp/supply-chain-management-vs-logistics.shtml

Jones, T. C., & Riley, D. W. (1985). Using inventory for competitive advantage through supply chain management. *International Journal of Physical Distribution & Materials Management*, *15*(5), 16–26. https://doi.org/10.1108/eb014615

Katina, P. F. (2016). Individual and societal risk (RiskIS): Beyond probability and consequence during Hurricane Katrina. In A. J. Masys (Ed.), *Disaster forensics: Understanding root cause and complex causality* (pp. 1–23). New York: Springer International Publishing. https://doi.org/10.1007/978-3-319-41849-0_1

Katina, P. F., & Keating, C. B. (2022). Deepwater horizon: Emergent behavior in a system of systems disaster. In L. B. Rainey & O. T. Holland (Eds.), *Emergent behavior in system of systems engineering: Real-World Applications* (pp. 193–230). London: CRC Press. https://doi.org/10.1201/9781003160816-11

Katina, P. F., Pinto, C. A., Bradley, J. M., & Hester, P. T. (2014). Interdependency-induced risk with applications to healthcare. *International Journal of Critical Infrastructure Protection*, *7*(1), 12–26. https://doi.org/10.1016/j.ijcip.2014.01.005

Keating, C. B., Katina, P. F., Chesterman, C. W., & Pyne, J. C. (Eds.). (2022). *Complex system governance: Theory and practice*. Springer International Publishing. https://link.springer.com/book/10.1007/978-3-030-93852-9

Kleindorfer, P. R., & Saad, G. H. (2005). Managing disruption risks in supply chains. *Production and Operations Management*, *14*(1), 53–68. https://doi.org/10.1111/j.1937-5956.2005.tb00009.x

Kopczak, L. R. (1997). Logistics partnerships and supply chain restructuring: Survey results from the US computer industry. *Production and Operations Management*, *6*(3), 226–247. https://doi.org/10.1111/j.1937-5956.1997.tb00428.x

Kumar, S., & Havey, T. (2013). Before and after disaster strikes: A relief supply chain decision support framework. *International Journal of Production Economics*, *145*(2), 613–629.

Lee, H. L., & Billington, C. (1992). Managing supply chain inventory: Pitfalls and opportunities. *MIT Sloan Management Review*, *33*(3), 65–73.

Lee, H. L., & Ng, S. M. (1997). Introduction to the special issue on global supply chain management. *Production and Operations Management*, *6*(3), 191–192. https://doi.org/10.1111/j.1937-5956.1997.tb00425.x

Li, S., Ragu-Nathan, B., Ragu-Nathan, T. S., & Rao, S. S. (2006). The impact of supply chain management practices on competitive advantage and organizational performance. *Omega*, *34*(2), 107–125.

Liang, X., Shetty, S., Tosh, D., Kamhoua, C., Kwiat, K., & Njilla, L. (2017). ProvChain: A blockchain-based data provenance architecture in cloud environment with enhanced privacy and availability. *Proceedings of the 17th IEEE/ACM International Symposium on Cluster, Cloud and Grid Computing*, 468–477. https://doi.org/10.1109/CCGRID.2017.8

Lummus, R. R., Krumwiede, D. W., & Vokurka, R. J. (2001). The relationship of logistics to supply chain management: Developing a common industry definition. *Industrial Management & Data Systems*, *101*(8), 426–432. https://doi.org/10.1108/02635570110406730

Lummus, R. R., & Vokurka, R. J. (1999). Defining supply chain management: A historical perspective and practical guidelines. *Industrial Management & Data Systems*, *99*(1), 11–17. https://doi.org/10.1108/02635579910243851

Luna, S., & Pennock, M. J. (2018). Social media applications and emergency management: A literature review and research agenda. *International Journal of Disaster Risk Reduction*, *28*, 565–577. https://doi.org/10.1016/j.ijdrr.2018.01.006

Majchrzak, A., Jarvenpaa, S. L., & Hollingshead, A. B. (2007). Coordinating expertise among emergent groups responding to disasters. *Organization Science, 18*(1), 147–161. https://doi.org/10.1287/orsc.1060.0228

Manoj, B. S., & Baker, A. H. (2007). Communication challenges in emergency response. *Communications of the ACM, 50*(3), 51–53. https://doi.org/10.1145/1226736.1226765

Mendonca, D., & Wallace, W. A. (2006). Impacts of the 2001 world trade center attack on New York City critical infrastructures. *Journal of Infrastructure Systems, 12*(4), 260–270. https://doi.org/10.1061/(ASCE)1076-0342(2006)12:4(260)

Mizutori, M., & Guha-Sapir, D. (2018). *Economic losses, poverty & disasters: 1998–2017.* United Nations Office for Disaster Risk Reduction. www.undrr.org/publication/economic-losses-poverty-disasters-1998-2017

Murayama, Y., Saito, Y., & Nishioka, D. (2013). Trust issues in disaster communications. *2013 46th Hawaii International Conference on System Sciences,* 335–342. https://doi.org/10.1109/HICSS.2013.576

Nahum, O. E., Hadas, Y., Rossi, R., Gastaldi, M., & Gecchele, G. (2017). Network design model with evacuation constraints under uncertainty. *Transportation Research Procedia, 22,* 489–498. https://doi.org/10.1016/j.trpro.2017.03.066

NASA. (2001). *Disaster data pathfinder.* Earthdata; Earth Science Data Systems, NASA. www.earthdata.nasa.gov/learn/pathfinders/disasters

NASA. (2007). *Systems Engineering handbook* (NASA/SP-2007–6105 Rev1; p. 360). National Aeronautics and Space Administration. www.nasa.gov/sites/default/files/atoms/files/nasa_systems_engineering_handbook.pdf

NOAA. (2021, March 11). *On this day: 2011 Tohoku earthquake and tsunami.* National Centers for Environmental Information (NCEI). www.ncei.noaa.gov/news/day-2011-japan-earthquake-and-tsunami

Ortuño, M. T., Tirado, G., & Vitoriano, B. (2011). A lexicographical goal programming based decision support system for logistics of Humanitarian Aid. *TOP, 19*(2), 464–479. https://doi.org/10.1007/s11750-010-0138-8

O'Sullivan, T. L., Kuziemsky, C. E., Toal-Sullivan, D., & Corneil, W. (2013). Unraveling the complexities of disaster management: A framework for critical social infrastructure to promote population health and resilience. *Social Science & Medicine (1982), 93,* 238–246. https://doi.org/10.1016/j.socscimed.2012.07.040

Papadopoulos, T., Gunasekaran, A., Dubey, R., Altay, N., Childe, S. J., & Fosso-Wamba, S. (2017). The role of Big Data in explaining disaster resilience in supply chains for sustainability. *Journal of Cleaner Production, 142,* 1108–1118. https://doi.org/10.1016/j.jclepro.2016.03.059

Paton, D., & Flin, R. (1999). Disaster stress: An emergency management perspective. *Disaster Prevention and Management, 8*(4), 261–267. https://doi.org/10.1108/09653569910283897

Perry, M. (2007). Natural disaster management planning: A study of logistics managers responding to the tsunami. *International Journal of Physical Distribution & Logistics Management, 37*(5), 409–433. https://doi.org/10.1108/09600030710758455

Petak, W. J. (1985). Emergency management: A challenge for public administration. *Public Administration Review, 45,* 3–7. https://doi.org/10.2307/3134992

Pujawan, I. N., Kurniati, N., & Wessiani, N. A. (2009). Supply chain management for disaster relief operations: Principles and case studies. *International Journal of Logistics Systems and Management, 5*(6), 679–692. https://doi.org/10.1504/IJLSM.2009.024797

Ransikarbum, K. (2015). *Disaster management cycle-based integrated humanitarian supply network management* [Dissertation, Clemson University].

Ransikarbum, K., & Mason, S. J. (2016). Multiple-objective analysis of integrated relief supply and network restoration in humanitarian logistics operations. *International Journal of Production Research, 54*(1), 49–68. https://doi.org/10.1080/00207543.2014.977458

Rasmussen, J., & Batstone, R. (1989). *Why do complex organisational systems fail?* World Bank Environmental Working Paper, No. 20. World Bank Environmental

Richey, R. G. (2009). The supply chain crisis and disaster pyramid: A theoretical framework for understanding preparedness and recovery. *International Journal of Physical Distribution & Logistics Management, 39*(7), 619–628. https://doi.org/10.1108/09600030910996288

Rinaldi, S. M. (2004). Modeling and simulating critical infrastructures and their interdependencies. *Proceedings of the 37th Hawaii International Conference on System Sciences*, 1–8. https://doi.org/10.1109/HICSS.2004.1265180

Rinaldi, S. M., Peerenboom, J., & Kelly, T. K. (2001). Identifying, understanding, and analyzing critical infrastructure interdependencies. *IEEE Control Systems, 21*(6), 11–25. https://doi.org/10.1109/37.969131

Ritchie, H., & Roser, M. (2014). *Natural disasters* (Version 2021) [Our World in Data]. Université catholique de Louvain. https://ourworldindata.org/natural-disasters

Sahebjamnia, N., Torabi, S. A., & Mansouri, S. A. (2017). A hybrid decision support system for managing humanitarian relief chains. *Decision Support Systems, 95*(C), 12–26. https://doi.org/10.1016/j.dss.2016.11.006

Samarajiva, R. (2005). Policy commentary: Mobilizing information and communications technologies for effective disaster warning: Lessons from the 2004 tsunami. *New Media & Society, 7*(6), 731–747. https://doi.org/10.1177/1461444805058159

Saunders, M. (1995). Chains, pipelines, networks and value stream: The role, nature and value of such metaphors in forming perceptions of the task of purchasing and supply management. *First Worldwide Research Symposium on Purchasing and Supply Chain Management*, 476–485.

Schipper, L., & Pelling, M. (2006). Disaster risk, climate change and international development: Scope for, and challenges to, integration. *Disasters, 30*(1), 19–38. https://doi.org/10.1111/j.1467-9523.2006.00304.x

Scholtens, A. (2008). Controlled collaboration in disaster and crisis management in the Netherlands: History and practice of an overestimated and underestimated concept. *Journal of Contingencies and Crisis Management, 16*(4), 195–207. https://doi.org/10.1111/j.1468-5973.2008.00550.x

Scholtens, K., Sharkey Scott, P., & Fynes, B. (2014). Mitigation processes: Antecedents for building supply chain resilience. *Supply Chain Management: An International Journal, 19*(2), 211–228. https://doi.org/10.1108/SCM-06-2013-0191

Shaluf, M. I. (2007). Disaster types. *Disaster Prevention and Management: An International Journal, 16*(5), 704–717. https://doi.org/10.1108/09653560710837019

Smith, W., & Dowell, J. (2000). A case study of co-ordinative decision-making in disaster management. *Ergonomics, 43*(8), 1153–1166. https://doi.org/10.1080/00140130050084923

Stumpe, F., & Katina, P. F. (2019). Multi-objective multi-customer project network: Visualising interdependencies and influences. *International Journal of System of Systems Engineering, 9*(2), 139. https://doi.org/10.1504/IJSSE.2019.100338

Swiss Re. (2021). *Global insured catastrophe losses rise to USD 112 billion in 2021, the fourth highest on record, Swiss Re Institute estimates.* www.swissre.com/media/press-release/nr-20211214-sigma-full-year-2021-preliminary-natcat-loss-estimates.html

Tan, K.-C., Kannan, V., & Handfield, R. (1998). Supply chain management: Supplier performance and firm performance. *International Journal of Purchasing & Materials Management, 34*(3), 2–9.

Tatham, P., & Christopher, M. (Eds.). (2014). *Humanitarian logistics: Meeting the challenge of preparing for and responding to disasters.* Kogan Page.

Tatham, P., & Kovács, G. (2010). The application of "swift trust" to humanitarian logistics. *International Journal of Production Economics, 126*(1), 35–45. https://doi.org/10.1016/j.ijpe.2009.10.006

Thomas, A., & Kopczak, L. R. (2005). From logistics to supply chain management: The path forward in the humanitarian sector. *Fritz Institute*, *15*, 1–15.

Tierney, K. (2012). Disaster governance: Social, political, and economic dimensions. *Annual Review of Environment and Resources*, *37*, 341–363. https://doi.org/10.1146/ANNUREV-ENVIRON-020911-095618

Tokgoz, B. E. (2012). *Probabilistic resilience quantification and visualization building performance to hurricane wind speeds* [PhD, Old Dominion University]. http://search.proquest.com.proxy.lib.odu.edu/pqdtlocal1005724/docview/1152026510/abstract/2766B19B3ACE4C73PQ/1

Tomasini, R. M., & Van Wassenhove, L. N. (2009). From preparedness to partnerships: Case study research on humanitarian logistics. *International Transactions in Operational Research*, *16*(5), 549–559. https://doi.org/10.1111/j.1475-3995.2009.00697.x

Vitoriano, B., Ortuño, T., & Tirado, G. (2009). HADS, a goal programming-based humanitarian aid distribution system. *Journal of Multi-Criteria Decision Analysis*, *16*(1–2), 55–64. https://doi.org/10.1002/mcda.439

Waugh, W. L., & Streib, G. (2006). Collaboration and leadership for effective emergency management. *Public Administration Review*, *66*(s1), 131–140. https://doi.org/10.1111/j.1540-6210.2006.00673.x

3 Developments in Disaster Management

3.1 DISASTER MANAGEMENT THEORIES

At this point, suffice it to say there are many challenges characterizing disaster management. A summary of these challenges revolve around (Cartier, 2009; de la Torre et al., 2012; Lin et al., 2011; Manoj & Baker, 2007; Paton & Flin, 1999) (i) accessibility to accurate information about the affected areas, (ii) limited communication, (iii) lack of efficient synchronization of different agents and transportation, and (iv) decisions load constraints. Clearly, *disaster management* is a vast domain of research. The present research focuses on the challenges of *disaster supply management*. Moreover, through this research, we propose a blockchain-enabled disaster supply chain and logistics management approach, underpinned by FEMA's qualified approach, which suggests that emergency management is a decentralized network of organizations collaborating to mitigate disaster impacts (Luna & Pennock, 2018).

Another challenge for disaster management is the use of theory in systematically planning and addressing disaster responses. The underlying theories (and heuristics) for a disaster situation, regardless of the context, are posited by Sementelli (2007). In this study, the aim is to create a basic taxonomy of current disaster theory. The range of theories is vast, from simple approaches to complex political elements. Sementelli's (2007) research suggests that disaster research can be categorized into two dimensions: concerns for tools and concerns for the process. Tools us on the process and naming with the focus on tools. Figure 3.1 depicts four associated categories: decision, administrative, social, and economical. Moreover, these theories can be used for several purposes, including the discovery of stressors as well as deep pathological (Katina, 2022) issues that can hinder disaster management. Decision theories have top-down approaches and are data-driven. Besides considering the theories, the other categories would focus on vulnerability, uncertainty, and ambiguity management of disasters (AliPour, 2021). Drabek (2007) also recognizes coordination concerns among the participants in a decentralization setting.

3.2 DISASTER SUPPLY CHAIN THEORIES

Theories play a significant role in attempts to make sense of complex environments (Keating et al., 2022; Tabaklar et al., 2015). In fact, there are calls for developing theories in the field of humanitarian logistics (Jahre et al., 2009). As the first attempt to theoretically support supply chain disaster preparedness and recovery, Richey (2009) argues that the focus on supply chains related to disasters has been expanded recently. It appears there is a scarcity of theoretical grounding in present studies. For example, Tabaklar et al. (2015) posit a lack of research on theoretical approaches toward

DOI: 10.1201/9781003336082-3

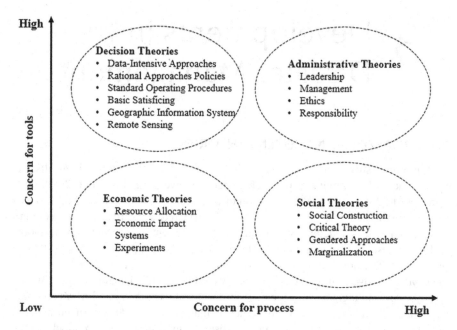

FIGURE 3.1 Categories of disaster theory, along with tools and process continuum.

humanitarian supply chain management. Their subsequent literature review provides a deeper understanding of the field from a theoretical view. The results indicate that humanitarian logistics is the primary discipline for the supply chain, and there is a lack of other theoretical perspectives and grand theories. However, Richey's (2009) research suggests that there are three keystones for disaster supply management— (i) competing values theory, (ii) collaborative view of the firm, and (iii) communication theory—all linked to resources management. The goal is to provide interconnection among stakeholders to fulfill the following aims:

- Obtaining effective partnerships and developing collaborative relationships for a long-term commitment
- Fostering information development and exchange to facilitate strategic planning
- Developing contingency programs for flexible response to the inevitable changes and following inconsistent goals

The performance and core objectives of humanitarian supply chain management are different from the traditional view of supply chain management. Nevertheless, the fundamental components are similar to other domains, including (i) supply chain network structure, (ii) supply chain business processes, and (ii) management components (Lambert et al., 1998; Tabaklar et al., 2015).

Assets play a vital role in disaster supply chain management. Assets can take different forms—human, financial, informational, digital (and technology), and physical assets. Moreover, we can use empirical (and theoretical) representations to manage assets (and other resources) while accounting for impact and interactions. Success might depend on how well resources match strategic initiatives in a disaster. In a humanitarian logistics strategy, the focus is on an innovative and responsive approach due to the highly dynamic nature of the field. Meanwhile, traditional logistics strategy focuses on improving efficiency and cost reduction (Jahre et al., 2009). Moreover, since disasters require short-term stabilization actions along with long-term recovery process strategies, issues of consequences demand (i.e., effort, investment, and time) are dynamic. Therefore, disaster supply chain and logistics management must proactively ensure redundancy to prevent the ripple effect. A disaster supply chain can be extraordinarily uncertain and dynamic, requiring unique management principles. Communication is one of the most critical factors in disaster relief operations due to the fast pace of disaster management. One approach that permits taking action is decentralization. Walmart undertook a similar approach to effectively provide food and water to the victims during the 2005 Hurricane Katrina disaster.

In the domain of humanitarian logistics, Jahre et al. (2009) recommend using both theoretical aspects of logistics and empirical aspects of humanitarian aid operations to improve human conditions. Moreover, three specific challenges (i.e., temporary network, vertical and horizontal coordination, and structure [centralized or decentralized]) affect disaster supply chain and logistics capabilities.

3.3 MODELING AND SIMULATION

Natural disasters have resulted in the mortality of almost three million people and affected the lives of 800 million people worldwide, which led to diseases and severe economic losses. Modeling and simulating the rescue procedure can help facilitate emergency management and limit the impact on society (Mustapha et al., 2013). Applying simulation as a tool to understand decision-making issues has recently gained considerable attention. The field of modeling and simulation (M&S), despite the label "emerging," is replete with several methods and tools for the analysis and synthesis of phenomena (Katina et al., 2020):

- *Agent-Based Modelling:* the emphasis is placed on "dynamically interacting, rule-based agents." These agents are created in software that defines possible agent actions, agent relations to other agents, and relations to the agent environment within a simulation (Hester & Tolk, 2010). ABM is described as a bottom-up modeling approach since the modeler is expected to model agent behavior at the individual level (i.e., item, part, component, and system), then global behavior emerges as a result of the interaction of agents in the environment (Borshchev & Filippov, 2004).
- *System Dynamics:* developed in the 1950s by Jay Forrester and is defined as a "study of information-feedback characteristics of industrial activity to show how organizational structure, amplification (in policies), and time delays (in decisions and actions) interact to influence the success of the enterprise" (Forrester, 1961). The need to understand "behavior and the

underlying structure of a complex system over time" is the major emphasis of SD. It can be described as a top-down approach based on nonlinearity and feedback control of internal processes and time delays that affect the system as a whole (Eusgeld et al., 2008), involving causal-loop diagrams and stocks-and-flow diagrams where a specific behavior (or parameter) such as X influences a second parameter Y and in turn Y influences X via feedback mechanism (Hester & Tolk, 2010).

- *Hybrid System Modeling:* hybrid system modeling (HSM) encompasses all mathematical modeling techniques that enable the modeling and simulation of "complex computational systems which display discrete and continuous system behavior" (Eusgeld et al., 2008, p. 21). HSM has a wide variety of applications in problem domains where creating detailed continuous models of (continuous) physical systems is, at the very least, impractical (Gomes et al., 2017). This can include modeling and simulation of transportation systems where discrete events control the flow and, as such, will tend to lean towards systems analysis (i.e., top-down approach), where a variety of interdependencies can be modeled.

- *Input-Output Model:* input-output model (IOM) was developed by Wassily Leontief and primarily focused on how changes in one sector (e.g., the economic sector) can affect another. IOM has frequently been used in economics to predict commodity or information flow between economic sectors. Several derived models exist, including the agent-based input/output inoperability model [AB-IIM].

- *Hierarchical Holographic Modelling:* hierarchical holographic modeling (HHM) attempts to view systems from various perspectives. HHM is grounded in the idea that understanding large-scale systems requires a multifarious approach involving (and not limited to) congeries of resources, objectives, actions, etc., which are non-commensurable and, at least potentially, conflicting elements (Haimes, 1981). In this case, multiple models can be developed to "capture the essence of many dimensions, visions, and perspectives of infrastructure systems" (Eusgeld et al., 2008, p. 25). This enables the discovery of infrastructure risks, vulnerabilities, and dependencies. Eusgeld et al. (2008) suggest that HHM is primarily a discovering process with applications ranging from military to water and transportation sector and, therefore, a top-down approach that can be used in different aspects of a system in question including interdependencies.

- *Critical Path Method:* critical path method (CPM) is a mathematically based engineering project management technique primarily used to schedule events (Kelley & Walker, 1959). Fundamentally, CPM involves (i) listing of all activities required to complete the project (typically categorized within a work breakdown structure); (ii) the time (duration) that each activity will take to complete; (iii) the dependencies between the activities; and (iv) logical endpoints such as milestones or deliverable items. Suppose one is interested in logical interdependencies among system inputs. In that case, CPM can be used to discover "logical interdependencies between activities, events, costs, and resources for process execution" (Eusgeld et al., 2008) and outputs. However, the availability of detailed data and information on the phenomena drive the usability of CPM.

- *Petri Nets:* Petri nets are often used in modeling standard modes of failures and cascading effects. An instantiation of Petri nets is stochastic Petri nets (SPN), which can be used to analyze the impact of communication on power grids (Schneider et al., 2006). Eusgeld et al. (2008) note that SPN is a "time enhanced variant of place-and transition nets, mathematical models of non-deterministic and discrete distributed systems. A Petri Net model is a bipartite-directed graph. It consists of places and transitions. Places may contain any number of tokens. When a transition switches ('fires'), it consumes the tokens from its input places, performs some processing task, and places a specified number of tokens into each of its output places" (Eusgeld et al., 2008, p. 33). In mapping infrastructure interdependencies, Rinaldi et al. (2001) use Petri nets to illustrate possible interdependencies among oil, transport systems, natural gas, electric power, water, and telecom systems.

The preceding section discusses M&S techniques primarily from "critical infrastructure" domain while emphasizing system interdependence and the ability to assess system vulnerability. Three primary and fundamental M&S paradigms stand out—agent-based, discrete-event, and system dynamics—and their value is undeniable. However, one could also argue that true validation of preparedness efforts comes from a response to an actual disaster event. Nevertheless, simulation can be used to evaluate the efficiency and effectiveness of the plan and the components in the preparedness phase. A controlled scenario-driven simulation can be designed to demonstrate and evaluate the capability to execute the operational tasks and procedures outlined in the contingency plan. The team of Ben Othman et al. (2017) is among the pioneers who research the case of supply chain management under emergency in cooperation with Airbus Defense and Space. Several studies have used simulation-based studies to plan evacuation operations (Hardy et al., 2008; Hardy et al., 2009; Pidd et al., 1996; Zou et al., 2005).

Modeling and simulation help in the fundamental analysis of the strength of a particular tool and the aims of a planned analytical endeavor (Hardy et al., 2009). Simulation can also be used to analyze the different types of disaster activities. Sebatli et al. (2017) provide a simulation-based approach to determine the supply requirements of an affected area from a disaster. Since a disaster is a complex phenomenon, there might be a need to apply different simulations to arrive at an optimum decision (Altay & Green, 2006; Bankes, 1993). Moreover, simulation can visualize supply chain risks effectively and address uncertainties (Aqlan & Lam, 2015). However, there aren't many frameworks that support the design and implementation of disaster simulation. In fact, limitations loom large when it comes to modeling (and simulation) in dealing with natural disasters:

- Nuances of natural disasters include organizational and structural dynamics related to behaviors associated with multiple roles.
- Not taking into account observable and indicators specific to the organization of the disaster, which the domain experts usually define.

- Issues in validation. Specifically, in evacuation models, validation refers to a systematic comparison of model predictions with sensitive information. However, there is a lack of relevant experimental data to feed the modeling can cause challenges.
- A weakness in representing subjects in simulations. Accurate occupants' representation according to comprehensive anthropometric data and human performance should be used to provide further validity to the model.
- There are limitations in the interoperability between emergency response modeling and simulation applications.
- Data transferring between emergency response simulation software applications is expensive.

When it comes to models addressing the previously mentioned issues, two frameworks are available: an agent-based disaster simulation environment and a dynamic discrete disaster decision simulation system. Agent-based disaster simulation environment provides model elements and tools to support the modeling and simulation of different types of disasters, describing how agents move, attach, and interact with each other and the environment (Mustapha et al., 2013). The discrete disaster decision simulation system is a comprehensive decision support system to simulate large-scale disaster responses (Mustapha et al., 2013). Multi-agent systems (MAS) are a powerful modeling tool for individual interaction simulation in a dynamic system. It has a distinctive ability to simulate situations that contains unpredictable behavior. MAS is one method used to model and simulate natural disaster emergencies, creating computer representations of dynamic events. The behavior of a set of entities can be modeled with a MAS. The application of MAS helps to experiment with all possible disaster scenarios and assists in decision-making (Mustapha et al., 2013).

At a general level, M&S presents an opportunity for use in low resource impact methods and tools, given the potentially high price associated with present realities especially realizing that simulations support decision-making in complex systems and help avoid costly mistakes. In fact, the 2006 National Science Foundation (NSF, 2006) suggests using simulation technology and methods to revolutionize engineering science because of the following reasons:

- Simulations are generally cheaper, safer, and sometimes more ethical than conducting real-world experiments.
- Simulations can often be even more realistic than traditional experiments as they allow the free configuration of environment parameters found in the operational application field of the final product. Examples are supporting the deepwater operation of the US Navy or simulating the surface of neighboring planets in preparation for NASA missions.
- Simulations can often be conducted faster than in real time. This allows using them for efficient if-then-else analyses of different alternatives, particularly when the necessary data to initialize the simulation can easily be obtained from operational data. This use of simulation adds decision support simulation systems to the toolbox of traditional decision support systems.

- Simulations allow for a coherent synthetic environment to integrate simulated systems in the early analysis phase via mixed virtual systems with first prototypical components to a virtual test environment for the final system. If managed correctly, the environment can be migrated from the development and test domain to the training and education domain in follow-on life cycle phases for the systems, including the option to train and optimize a virtual twin of the real system under realistic constraints even before first components are being built.

Modeling and simulation can also study disruption propagation and the ripple effect across multiple systems and allow for an in-depth view of network operations, even in real-time operation situations providing means to capture behavior changes. For disaster management, simulations are relevant. A simulation model can analyze supply chain risks and help develop contingency plans and efficient disaster management. Since supply chains can face all sorts of uncertainties (e.g., demands, change in processes), Swaminathan et al. (1998) suggest applying simulation to deal with the uncertainties. In addition, simulation can enable the decision-makers to quantitatively assess the risks as well as opportunities associated with supply chain options. In fact, modeling and simulation is commonly used for planning evacuation operations, analyzing the different types of disaster operation management, including resource allocation and mass decontamination, and relief supplies distribution operations (Rodrigo et al., 2018).

3.4   UBIQUITOUS TECHNOLOGIES

Information is vital for effective disaster management. Information systems extensively record, exchange, and process information (Sakurai & Murayama, 2019). Even though the new information technologies have changed situational information collection, Sakurai and Murayama (2019) argue that there is a scarcity of how technologies are implemented within disaster strategies. Nevertheless, information technologies can be applied in each phase of disaster management (Sakurai & Murayama, 2019):

- *Phase 1: Mitigation*—information technology can be used for real-time monitoring of risk. Examples include sensor network systems (vulnerability of infrastructures) and unmanned aerial vehicles.
- *Phase 2: Preparedness*—information technology can be used for living-lab scenario-simulation (e.g., virtual reality), messenger application and online dashboards, and knowledge repository based on previous disaster experiences.
- *Phase 3: Response*—information technology can be used to provide situational information as in situational awareness (social networks) and enable decision-making.
- *Phase 4: Recovery*—information technology can be used to manage recovery operations to streamline coordination of available resources, share victim data, provide interactive communication, and enhance collaboration with disaster relief agencies.

A general disaster plan explains the chain of command in a disaster duration. However, no holistic strategy indicates who should use what technology for what reason (Sakurai & Murayama, 2019). And yet, the high frequency of disasters is a continuing fact of the lack of up-to-date technology systems for organizations that must address disasters. Moreover, the main challenge of humanitarian logistics networks is the requirement for better coordination.

3.4.1 Cloud Computing

Cloud computing is the on-demand availability of computer system resources, especially data storage (cloud storage) and computing power, without direct active management by the user. (Montazerolghaem et al., 2020; Ray, 2018). The National Institute of Standards and Technology's definition of cloud computing identifies five essential characteristics (Mell & Grance, 2011):

- *On-Demand Self-Service:* A consumer can automatically use unilaterally provision computing capabilities, such as server time and network storage, without requiring human interaction with each service provider.
- *Broad Network Access:* Capabilities are available over the network and accessed via standard mechanisms that promote use by heterogeneous thin or thick client platforms (e.g., mobile phones, tablets, laptops, and workstations).
- *Resource Pooling:* The provider's computing resources are pooled to serve multiple consumers using a multi-tenant model. Different physical and virtual resources are dynamically assigned and reassigned according to consumer demand.
- *Rapid Elasticity:* Capabilities can be elastically provisioned and released, in some cases automatically, to scale rapidly outward and inward commensurate with demand. To the consumer, the capabilities available for provisioning often appear unlimited and can be appropriated in any quantity at any time.
- *Measured Service:* Cloud systems automatically control and optimize resource use by leveraging a metering capability at some level of abstraction appropriate to the type of service (e.g., storage, processing, bandwidth, and active user accounts). Resource usage can be monitored, controlled, and reported, providing transparency for both the provider and consumer of the utilized service.

Cloud computing has emerged as a technology and has the potential to revolutionize the information and communication landscape. Moreover, cloud computing allows the development of reliable, agile, and incrementally deployable and scalable systems at low cost and access to large shared resources on demand. Alazawi et al. (2011) proposed an intelligent disaster management system. The intelligent system can gather information from multiple sources and locations, make effective strategies and decisions, and propagate the information to other nodes in real time. Social media is the new frontier source of information for disaster relief agencies, improving situational awareness and two-way communication (Sakurai & Murayama, 2019).

3.4.2 BLOCKCHAIN TECHNOLOGY

Blockchain is considered the fifth disruptive innovation following mainframe, personal computer, internet, mobile, and social network, and it is the fourth milestone of the credit evolution. Blockchain is predicted to remodel human society drastically and will promote the present information internet to the credit internet (R. Xu et al., 2017). Harber and Stornetta incepted the basic idea of blockchain in 1991. The first version was created in 2008 with an original white paper on Bitcoin by Satoshi Nakamoto published on his internet page (Zile & Strazdina, 2018). Cryptocurrency Bitcoin was the first practical solution for applying blockchain technology and one of the main reasons for blockchain's current popularity. Based on the World Economic Forum report in 2015, blockchain has been considered one of the megatrends that will change the world in the next decade (Kshetri, 2018). Again, blockchain technology is formally defined as a fully distributed system for cryptographically capturing and storing a consistent, immutable, linear event log for transactions between network actors (Risius & Spohrer, 2017. Transparency is enforced within the network with a system-wide consensus on the validity of the entire history of transactions (Queiroz et al., 2019).

Blockchain consists of nodes within a communication network that contains a common communication protocol. Each node stores a copy of the blockchain on the network, and a consensus function verifies transactions to preserve the immutability of the chain (Guo & Yu, 2022). Each block is identified via its cryptographic hash. Each block is referred to the previous block's hash, which creates a link between blocks that form a blockchain. The transactions of each block are hashed in a Merkle tree. The root hash and the previous block's hash are recorded in the block header. Blockchain provides interaction between users using a pair of public and private keys (Casado-Vara et al., 2018). The hashing process transforms assets into a digitally encoded token that can be registered, tracked, and traded with a private key on the blockchain (Ivanov et al., 2019). Every blockchain network needs a distributed consensus mechanism. As the network approves the transaction, it will be a valid and permanent part of the database. This method can significantly enhance the traceability, transparency, and trust of the entire system (Tian, 2017). Users have access to the audit trail of activity. The decentralized data storage decreases any single point's failure risk (Guo & Yu, 2022). Figure 3.2 depicts blockchain operations adapted from Yoo and Won (2018).

Research also suggests that the creation and movement of digital assets, decentralization, disintermediation, immutability, tamperproofing, and transaction sharing

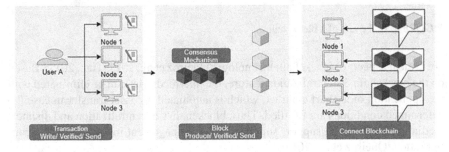

FIGURE 3.2 Operations of a blockchain.

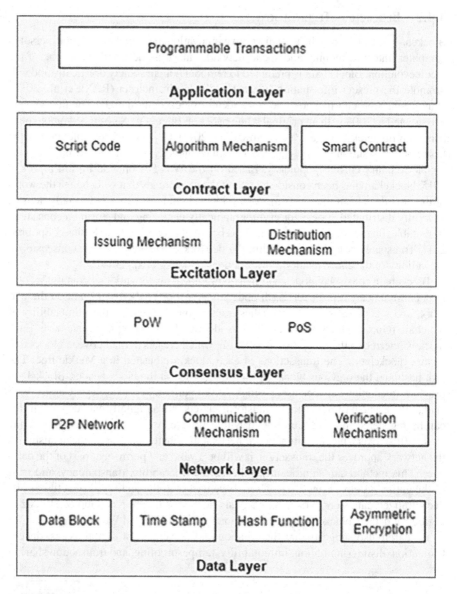

FIGURE 3.3 Layers of a blockchain.

are critical aspects of blockchain technology (Queiroz et al., 2019; Tian, 2017). Based on the blockchain decentralization feature, the intermediaries can be eliminated with the application of a smart contract, which is automated for asset transfer in case the determined conditions are fulfilled. Thus, blockchain's decentralization and disintermediation features can support supply network management innovation and reconfiguration (Queiroz et al., 2019).

Figure 3.3 illustrates six layers of a blockchain system (R. Xu et al., 2017). A fundamental feature of blockchain-based networks is decentralization. And decentralization is essential in key facts of the disaster supply chain, namely *distributed trust* and *consensus*. Hence, large networks have the integrity of transactions among the peers in a peer-to-peer configuration without the central mediate third party. With verifiable trust, the networks can be audited in a trusted and transparent manner. Blockchain facilitates operations that require interactions among several stakeholders providing transparency and trust without involving any third party. In fact, Hassan et al. (2019) posit that the immutability, transparency, and peer-to-peer consensus features of blockchain provide a trusted audit of networked systems. In the meantime, the control is maintained on the edge of a network of chains.

Transactions among the peers are stored as a record in a data structured chain series, and every network member preserves a copy of this record (Hassan et al., 2019). A hash is a unidirectional cryptographic function, and it takes an arbitrary input of an arbitrary length that generates a random fixed-length string of characters. Hence, each output is unique and footprint for the input. Hash can be used to check the integrity of a piece of data. Merkle trees are one of the essential components of a blockchain that support robust blockchain functionality and enable efficient and secure verification of large data structures. The Merkle tree structure is hash-based, which can ensure data integrity. With any changes, the hashes on the path from the root to the changing leaves are changed (Weber et al., 2019). Figure 3.4 depicts the four main fields for each block (Hassan et al., 2019). The first block in a blockchain is called the genesis block. The contents of a hash value include block number, previous hash, and data recorded in the hash field. The previous hash is an essential part that contains the value of the previous block hash. If the contents of a block change, it would be reflected in the hash of the respective block. The changes are also reflected in the portion of the block that comes next within the blockchain. Hence, the records stored in a blockchain are immutable since hashing and the distribution of the blockchain are copied among the network's peers (Hassan et al., 2019).

For each block to be added to the chain, all network participants must confirm the authenticity, which means reaching the network consensus and being validated by all the nodes to the point that all the nodes have an up-to-date blockchain structure (Zile & Strazdina, 2018). The peers of the blockchain network are informed

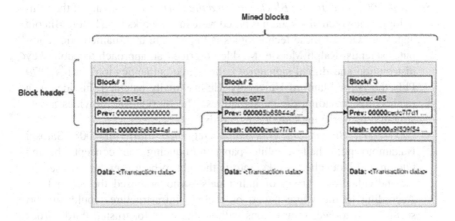

FIGURE 3.4 An example of a blockchain mined block.

of the overall state of the stored records in the blocks using the consensus engines. Consensus engines include proof-of-authority (PoA), proof-of-stake (PoS), and proof-of-work (PoW). These coordinate the nodes and decide which block should be added to the ledger next (Lohmer et al., 2020). The main goal of the consensus mechanism is to provide a verifiable trust guarantee (nonce) to prevent double-spending the digital asset. Finding a nonce is a computationally rigorous process called mining. A nonce is an integer that produces a hash matching a predefined pattern when hashed together with the contents of a block. As the miner finds a nonce, the network rewards the node with a set number of cryptocurrency tokens (Hassan et al., 2019).

Noteworthily, blockchain technology provides global, tamper-resistant, and append-only ledgers for a namespace system that guarantees the integrity, availability, uniqueness, and security of name-value pairs. It delivers the necessary fundamentals for conducting secure and distributed naming services (Hassan et al., 2019). The intended application of blockchain and the restrictions, including scalability and accessibility, determine the mechanism of achieving immutability. Again, access to transactions can vary based on mechanisms for validating transactions: permissioned and permissionless. In public, all nodes can read and submit transactions. Only authorized nodes can validate transactions (i.e., permissioned). On the other hand, under public-permissionless, all nodes can read, submit, and validate transactions. In private permissioned chains, only authorized nodes can read, submit, and validate transactions. And in consortium permissioned chains, the chain is partially decentralized and controlled by a group of organizations that have the authority to participate by running a full node and mining. Data distribution is the other important part of blockchain technology (Zile & Strazdina,2018). Permissioned and private blockchains also are available in addition to the public ones.

Linux Foundation's Hyperledger Fabric (HLF) popularized this concept for business use cases that requires confidentially in addition to data immutability and peer-to-peer consensus. These two types deploy a cryptographic membership service on top of the immutable record-keeping of the blockchain. Hence, each member can be identified based on the real-world identity (Hassan et al., 2019). The following timeline provides further elaboration on the evolving nature of blockchain technology (Sheldon, 2021):

- *1979–2007: Creation of Blockchain and the Early Years*—many of the technologies on which blockchain is based were in the works long before Bitcoin appeared. One of these technologies is the Merkle tree, named after computer scientist Ralph Merkle. Merkle described an approach to public-key distribution and digital signatures called "tree authentication" in his 1979 PhD thesis for Stanford University. He eventually patented this idea as a method for providing digital signatures. The Merkle tree provides a data structure for verifying individual records.
- *2008–2009: Bitcoin and Blockchain Get Their Start*—in 2008, Satoshi Nakamoto published a white paper introducing the concepts behind bitcoin and blockchain. Nakamoto is thought to be a pseudonym used by the individual—or group of individuals—who proposed the technology. According to the white paper, blockchain infrastructure would support secure, peer-to-peer transactions without the need for trusted third parties

such as banks or governments. Nakamoto's true identity remains a mystery, but there has been no shortage of theories.

- *January 3, 2009*—Nakamoto mined the first bitcoin block, validating the blockchain concept. The block contained 50 bitcoins and was known as the Genesis block—aka block 0.
- *January 8, 2009*—Nakamoto released Bitcoin v0.1 to SourceForge as open-source software. Bitcoin is now on GitHub.
- *January 12, 2009*—the first bitcoin transaction took place when Nakamoto sent Hal Finney 10 bitcoin in block 170.
- *October 12, 2009*—the #bitcoin-dev channel was created on Internet Relay Chat for bitcoin developers.
- *October 31, 2009*—the first bitcoin exchange, Bitcoin Market, was established, enabling people to exchange paper money for bitcoin.
- *November 22, 2009*—Nakamoto launched the Bitcointalk forum to share bitcoin-related news and information.
- *2010–2012: Bitcoin and Cyber Currency Take Hold*—on May 22, 2010, bitcoin made history when Laszlo Hanyecz paid 10,000 bitcoin for two Papa John's pizzas. The pizzas were valued at around $25, a trade that would now be worth more than $350 million.
- *2013–2015: Ethereum and Blockchain Rise to Fame*—when 2013 arrived, Bitcoin was well-established and continued on its upward trajectory. In February, Coinbase reported selling $1 million worth of bitcoin in a single month at more than $22 each. By the end of March, with 11 million bitcoin in circulation, the currency's total value exceeded $1 billion. And in October of that year, the first bitcoin ATM was launched in Vancouver, BC.
- *2016–Present: Blockchain Goes Mainstream*—today, many industries view blockchain as a valuable technology, separate from Bitcoin or other cyber currencies. Despite this trend, each year from 2016 to the present had its ups and downs.
  - 2016
    - The term "blockchain" gained acceptance as a single word rather than being treated as two concepts, as they were in Nakamoto's original paper.
    - The Chamber of Digital Commerce and the Hyperledger project announced a partnership to strengthen industry advocacy and education.
    - A bug in the Ethereum decentralized autonomous organization code was exploited, resulting in a hard fork of the Ethereum network.
    - The Bitfinex bitcoin exchange was hacked, and nearly 120,000 bitcoin were stolen—a bounty worth approximately $66 million.
  - 2017
    - Bitcoin hit a record high of nearly $20,000.
    - Japan recognized Bitcoin as a legal currency.
    - Seven European banks formed the Digital Trade Chain consortium to develop a trade finance platform based on blockchain.
    - Block.one company introduced the EOS blockchain operating system, designed to support decentralized commercial applications.

- Approximately 15% of global banks used blockchain technology in some capacity.
- 2018
  - Bitcoin turned ten this year.
  - Bitcoin value continued to drop, ending the year at about $3,800.
  - The online payment firm Stripe stopped accepting Bitcoin payments.
  - Google, Twitter, and Facebook banned cryptocurrency advertising.
  - South Korea banned anonymous cryptocurrency trading but announced it would invest millions in blockchain initiatives.
  - The European Commission launched the Blockchain Observatory and Forum.
  - Baidu introduced its blockchain-as-a-service platform.
- 2019
  - Walmart launched a supply chain system based on the Hyperledger platform.
  - Amazon announced the general availability of its Amazon Managed Blockchain service on AWS.
  - Ethereum network transactions exceeded 1 million per day.
  - Blockchain research and development took center stage as organizations embraced blockchain technology and decentralized applications for a variety of use cases.
- 2020
  - Nearly 40% of respondents incorporated blockchain into production, and 55% viewed blockchain as a top strategic priority, according to Deloitte's 2020 Global Blockchain Survey.
  - Ethereum launched the Beacon Chain in preparation for Ethereum 2.0.
  - Stablecoins saw a significant rise because they promised more stability than traditional cybercurrencies.
  - There was a growing interest in combining blockchain with AI to optimize business processes.

A permissionless public blockchain relies on decentralization, immutability, and anonymity. As such, it eliminates the need for a third party. It is open for participation for transacting or validating, higher transaction speed, and lower fees. A permissioned blockchain relies more on efficient consensus algorithms, and the value of the stored data depends on the unambiguity (Ziolkowski et al., 2018).

Several blockchain applications have been introduced: blockchain 1.0 for cryptocurrencies; blockchain 2.0 for economic, market, and financial applications; and blockchain 3.0 for applications beyond currency, finance, and markets (Rodrigo et al., 2018). Blockchain technology can be the key to the essential ingredients to confront the content distribution challenges and a more agile method for content delivery with a more trusted, autonomous, and intelligent network (Hassan et al., 2019). Blockchain can be a breakthrough discovery that changes daily activities and processes in different disaster supply chain applications. For those interested in an extensive set of famous blockchain use cases, ranging from lottery to marriage registration, we suggest Zile and Strazdina (2018). The most frequent application efforts of blockchain are in the financial sector and then government, insurance, and healthcare services. However, blockchain technology is still at the initial stage of adaptation.

Blockchain evaluation models are also readily available. For example, a multilevel framework can be applied to evaluate the initial suitability of blockchain technology (Verma et al., 2019). In this case, the framework evaluates different technical aspects of blockchain and provides a step-by-step means to decide on the application of blockchain technology. Applying the framework in different use cases claims blockchain is suitable for supply chain and identity data management projects (Zile & Strazdina, 2018). Effective supply chain management is challenging in every sector, but disaster management has added complexity and risk due to its direct impact on victims' lives. The ability of blockchain to receive and record a massive amount of data and real-time availability for every part of the process provides one of the most transparent and secure ways to deal with data and enable observing the online activities of the project (Akram & Bross, 2018). Most research focuses on designing blockchain-based systems to achieve traceability by leveraging the main properties of the blockchain (Weber et al., 2019). The capacity to utilize smart contracts to automate processes and reduce costs is a crucial mechanism that blockchain technology could support supply chain performance enhancement.

It is also recognized that blockchain technology can enhance supply chain management by reducing (eliminating) fraud and errors, reducing delays from paperwork, improving inventory management, identifying issues more rapidly, minimizing costs, and increasing consumer and partner trust (Clauson et al., 2018). One of the main differences between blockchain and other existing technologies is the lower costs of adding new participants, data encryption, and record validation. Blockchain provides a faster transaction by reducing the required time of obtaining confirmation from multiple participants, providing reliable and verified information, and allowing automation of some transaction logic through smart contracts (Babich & Hilary, 2018).

There are also prospective benefits of automation through blockchain technology. It can increase coordination throughout the supply chain. If the machine's downtime is unavoidable, the customer would be informed of the potential disruption to their supply. Blockchain technology with smart contracts permits the automatic activation of the entire supply chain. The critical challenges in disasters could be addressed with superior supply chain management practices that are digitally enabled by blockchain technology (Clauson et al., 2018). A blockchain-based disaster supply chain system can enhance communication among stakeholders to receive real-time feedback and responses. Multi-signature governance accounts provide more security by distributing transaction approval to crucial decision-makers that are involved in the network. Applying blockchain to the disaster response process can improve the collection and sharing of data processes. The ordering and tracking updating information will be linked in a chain that utilizes a peer-to-peer validation process to build trust within the system. All the valid network participants can view the changes and updates, hence assisting in improving transparency and transaction audit (Anand et al., 2017). Moreover, blockchains can be practical in disaster relief processes by decreasing bureaucracy and reporting requirements, providing better decision-makers access to quicker fund availability and long-term funding. Stakeholder voting can be implemented based on the blockchain's distributed consensus authentication and irreversibility feature (R. Xu et al., 2017).

Moreover, blockchain technology has been applied in the disaster supply chain and logistics management. While an in-depth discussion is provided in Section 3.5.1, the research of Clauson et al. (2018) provides an overview of opportunities (and challenges) associated with blockchain technology adoption as well as deployment in the health supply chain. Moreover, Hassan et al. (2019) posit blockchain-based solutions for finding common patterns, differences, and technical limitations in decision-making. Nevertheless, while some studies look at the intersection of blockchain technology and supply chain management, there is still a scarcity in exploring how blockchain technology can be utilized to enhance disaster supply chain and logistics networks, especially considering the importance of communications.

3.4.3 Smart Contracts

A smart contract is simply a computer program or a transaction protocol that is intended to automatically execute, control, or document legally relevant events and actions according to the terms of a contract or an agreement (Casado-Vara et al., 2018). Nick Szabo introduced the smart contract concept in 1994, where the computerized transaction protocol executes the terms of a contract (Kushwaha & Joshi, 2021), intending to reduce the costs and delays of traditional contracts and satisfy common contractual conditions. Smart contracts are considered an algorithmic enforcement of an agreement among mutually non-trusting entities. The smart contract includes functions that receive input parameters of the contract and get invoked when the transactions are made. A second is the state variable dependent on the logic developed in the functions. Compilers convert the written programs into bytecode and deploy them to the blockchain network. An arbitrary value is transferred in an immutable manner that the conditional transactions are recorded, executed, and distributed across the blockchain network (Hassan et al., 2019). The transaction in the smart contract is executed independently and automatically in a prescribed way on every node of the network based on the data included in the triggered transaction (Casado-Vara et al., 2018). Smart contracts decrease the legal and enforcement expenses and the regulating authority requirement. It creates a trusted environment for the members from multiple contrasting and diverse communities (Hassan et al., 2019). The use of a smart contract in blockchain technology enables the creation of a self-governing collaboration with enforceable rules of interactions without the need for a central governing body (Ziolkowski et al., 2018). The use of blockchain technology in disaster management has attractive benefits (AliPour, 2021):

- *Distributed Funding:* payments can be made to service providers based on the smart contract achievements. All share accountability for overruns and payment is based upon the performance of agreed-upon requirements.
- *Multi-Party Validation:* all participants can validate the parameters in the progress of the project. Then the oversight board would increase the effectiveness and value of performing an audit.

- *Smart Contract Functionality:* disaster management policy can be scripted with the smart contract to provide the disaster's requirements, enabling all stakeholders to approve milestones. Damage can be logged, and costs can be estimated early in the recovery process.

Moreover, smart contracts can also ensure data security and confidence in data quality (Akram & Bross, 2018). This means that information collection would be more accessible via blockchain technology when dealing with multiple organizations. The aggregation of information can be done selectively, and just the required information for improving system efficiency is shared. The aggregated information can be used synergistically to improve the system's performance and reallocate benefits among participants. Smart contracts can automate actions based on aggregate information, decrease the lead time, and generate markets in which non-standardized resources are traded (Babich & Hilary, 2018).

Additionally, the decision-making process of disaster response varies significantly from the conventional decision-making process. In a disaster response, challenges of volatility, uncertainty, complexity, and ambiguity take the forefront. Moreover, the disaster may be uncontrollable, the time leading to decision-making may be limited and yet available information may be unreliable information, and yet the decision may have lasting implications (Altay & Green, 2006). Therefore, information transmission among all the parties during disaster phases plays a significant role in achieving the objectives. Creating a standard operational structure is necessary for communication as well as coordination of activities among the responsive organizations, which requires an appropriate level of shared information and participation in the disaster operation at multiple locations. All actors should be aware of the limits in the possible combination of collaborations and support under a set of conditions.

Moreover, one can expect challenges when more diverse organizations are included in the disaster operations. Communication in emergency management includes the capacity to build shared meanings among the involved participants, while the focus has been more on the interoperability of the mechanical devices (e.g., radio, cellphone). This process requires the capacity of resonance between the organization and the environment for innovation or new ways to solve the problem (Comfort, 2007). A white house homeland security report based on the lesson learned notes, "The lack of communication and situational awareness had a debilitating effect on the federal response" (Townsend, 2006, p. 50). Adding cognition to emergency management would include systematic means of adapting to the dynamic and uncertain conditions of a disaster evolvement.

Coordination depends on communication and control. In this case, control in emergency management is the capacity to keep actions focused on the mutual objectives of saving lives and properties and maintaining continuous operations. This requires maintaining shared knowledge, skills, and mutual adjustment of actions to suit the needs of the dynamic situation (Comfort, 2007). These capacities have relied on a well-designed information structure that facilitates communication, coordination, control, and cognition among the participating actors in emergency response (Comfort, 2007). Information structure is fundamental since it can construct a human capacity to learn and use technology to monitor performance, facilitate detection and correction of errors, and increase creative problem-solving capability (Comfort, 2007).

3.4.4   IoT: INTERNET OF THINGS AND 6G

The Internet of Things (IoT) describes the network of physical objects (i.e., "things") embedded with sensors, software, and other technologies to connect and exchange data with other devices and systems over the internet. While IoT promises benefits for society, IoT security "has not kept up with the rapid pace of innovation and deployment, creating substantial safety and economic risks" (US DHS, 2016, p. 2). In the future, where the sheer volume of things connected to the internet is projected to increase and involve such systems as automated vehicles and industrial systems, there will be a need for low latency and high speed in the distributed computing systems to save time and bandwidth. Interestingly, according to Cisco Annual Internet Report (2018–2023), IoT devices will account for 50% (14.7 billion) of all globally networked devices by 2023. Device manufacturers, business intelligence software firms, mobile carriers, systems integrators, and infrastructure vendors will have unique but complementary roles across the IoT landscape (Cisco, 2020).

The IoT landscape is ripe with challenges and opportunities. These challenges and opportunities include addressing users (and devices) and connections, especially in addressing growth in internet users, shifts in the mix of devices and connections, machine-to-machine (M2M) applications across many industries, accelerating IoT growth, and the rise in mobility. Second is addressing network performance and user experience, especially in Wi-Fi, the effects of accelerating speeds on traffic growth, and tiered pricing/security analysis. Third is multi-domain architecture in which society must reimagine applications, transform infrastructure to reduce costs, address the security of devices (and connections, networks, and data), and empower employees and teams (Cisco, 2020). The emerging sixth-generation (6G) standard currently under development for wireless communications supporting cellular data networks is poised to address many of these issues. 6G is a planned successor to 5G and will likely be significantly faster (Fisher, 2022; US DHS, 2022).

IoT provides new services to improve daily life where big data, cloud computing, and monitoring can take part. A wireless sensor network is a subset consisting of small sensing devices with few resources wirelessly connected. Nodes can communicate with other internet-enabled devices using sensors. Sensory features collect information from the environment through specific sensors and then process and transmit it to the internet (Plageras et al., 2018). The IoT-based sensors provide efficient monitoring of the processes. The IoT-generated sensor data has the characteristics of real-time, large amounts, and unstructured type (Syafrudin et al., 2018).

IoT technology can play a critical role in developing adequate monitoring infrastructure systems as well as enhance information sharing in disaster management systems. Innovative real-time monitoring and disaster warning systems have been developed based on the IoT paradigm, where objects are globally interconnected. Real-time monitoring and disaster warning systems are part of early warning systems, which are capacities needed to generate and disseminate timely and meaningful warning information to enable individuals, communities, and organizations threatened by a hazard to prepare and act appropriately and in sufficient time to reduce the possibility of harm or loss. Wireless sensors network (WSN) is a part of IoT that has been widely used for monitoring natural disasters in remote and inaccessible areas. WSNs are based on autonomous sensor nodes with low-energy monitoring and recording capability of the surrounding environment (Adeel et al., 2019).

3.5 DYNAMIC VOLTAGE FREQUENCY SCALING (DVFS) ALGORITHM

Dynamic voltage and frequency scaling (DVFS) was proposed by Gu et al. (2014). DVFS is a sorting process embedded in the server, enabling analyzing the disaster management complexity. DVFS is a tool that can be used to analyze the maximum level of disaster management complexity. An awkward request from one of the network nodes, which can be an army of clients in case of a disaster, can be detected with the sorting feature of DVFS. The attacked host within the network can be identified. The attacks from the customers can be stored on the server and managed by the disaster based on real-time information. By using local and external algorithms, the stored information can be linked to it, enabling the management of the enforced traffic on the network from the disaster node. Employing DVFS can support managing the vast input data from the disaster and the processing costs. By sorting data on servers and evaluating the transaction power, the nodes on which the disaster occurred can be identified due to the high rate of data added to that node. The information will be rewritten once in the token-based blocks in the network. When the blocks on the disaster location servers are rewritten, the monitoring cost will decrease to zero.

The concept of DVFS has been applied in the literature to reduce the proposed applications' energy consumption. For example, Calheiros and Buyya (2014) targeted the energy-efficient execution of cloud models in disaster management. They applied DVFS to enable deadlines for executing urgent CPU-intensive jobs with less energy consumption. Furthermore, Hosseini Shirvani et al. (2020) emphasized the need for studying virtual machine migration and DVFS techniques on disaster-resilient fault-tolerant systems.

3.5.1 Systematic Literature Review

This section describes literature pertaining to disaster supply chain and logistics networks. Figure 3.5 depicts the schema used in the literature review. The review focused on three topics: blockchain supply chain, disaster management, and humanitarian logistics.

In literature, numerous endeavors have been done to analyze and visualize many different types of bibliometric networks. The most frequently studied types of bibliometric networks are based on citation relations. At this point, we emphasize the need for (and value of) direct citation networks to study the history and development of research fields. Algorithmic historiography, which ultimately led to the program HistCite™, is a flexible software solution to aid researchers in visualizing the results

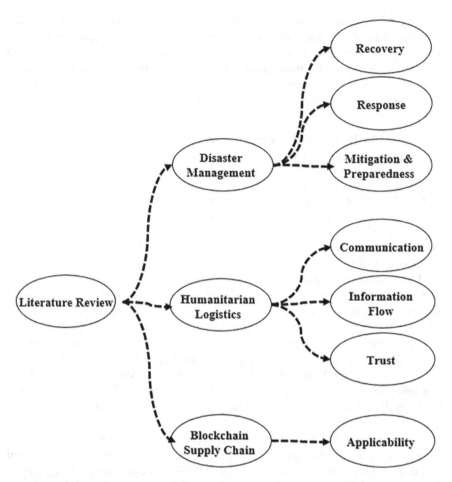

FIGURE 3.5 The schema of literature review.

of literature searches on the Web of Science. It is easy, fast, and provides perspectives and information unavailable on the Web of Science (Mijnhardt, 2007). The scope of the literature review included analysis items and category values. Each analysis item (i.e., disaster phases, application domain, impact on disaster, impact on victims, assessment layers, data collection, phases of the study, method of the study) corresponds to a category value (i.e., preparedness, planning, response, recovery, emergency management, response pace and quality, equity, fairness, communication, four phases of a disaster, historical data, simulation data, simulation, case study, algorithmic historiography).

Java-based software tool for analyzing and visualizing direct citation networks (van Eck & Waltman, 2014) is then applied. The software offers sophisticated functionality for drilling down into citation networks dealing with a specific topic of interest. A literature review discusses published information in a particular subject area—sometimes information in a particular subject area within a certain time period.

A literature review can be just a simple summary of the sources, but it usually has an organizational pattern and combines both summary and synthesis. A summary is a recap of the vital information of the source, but a synthesis is a re-organization, or a reshuffling, of that information. It might give a new interpretation of old material or combine new with old interpretations. Or it might trace the intellectual progression of the field, including major debates. And depending on the situation, the literature review may evaluate the sources and advise the reader on the most pertinent or relevant (Etemadi et al., 2021; The Writing Center, 2022). The disaster supply network articles are reviewed in the first phase, focusing on disaster network communication. The principal components of the disaster supply chain within the literature included analysis items (partnerships, strategic networks, supply network design, distribution base integration, contract view, communication, knowledge transfer) and their corresponding categorical values (trust, commitment, partnership performance, information flow, time compression, distribution channel management, organizational structure, technology transfer).

The Web of Science database is used for keyword searches, including the combinations shown in Table 3.1. The procedures followed for this systematic literature review include determining the Web of Science database as the primary research database from 1999 to 2021. Also, the English language within journal publications was considered for this review. Web of Science is a human-curated database (Kendall, 2019), and as such, the following apply:

- Journals are the focus of the Web of Science, and they are selected for inclusion by humans based on scholarly and quality criteria by literature review committees.
- Data about each article is entered into the database uniformly: author, title, date, journal name. This means you get accurate retrieval when searching for those things. Results can be sorted reliably by the latest date.
- Articles on the Web of Science are tagged with important information about their structure, such as "review article." They are not tagged for content, so you must include all possible variations of the topic you are searching.

TABLE 3.1

Results of Literature Review Based on Selected Keywords in Web of Science

Search Query (Article) Language (English)	Web of Science Search Result (1999–2021)	Sum of Times Cited
"Disaster" AND "Supply Chain"	618	10883
"Disaster" AND "Supply Network"	44	759
"Disaster" AND "Blockchain"	16	148
"Blockchain" AND "Humanitarian"	14	59
"Blockchain" AND ("Disaster" AND "Supply Chain"	7	85

- These databases have subject specificity in that journals are chosen for inclusion in the Web of Science based on the subject matter. Web of Science is interdisciplinary and includes each subject area's "best" journals.

Accurate retrieval means that search results are reproducible and reportable (significant for systematic reviews).

The citation network in CiteNetExplorer is acyclic due to the visualization of the citation flow. The CiteNetExplorer is applied to find the citation mapping system for the influential papers to find the roots and paths of methods related to the disaster supply network.

The initial search for "disaster supply chain" provides a visual tree map of the search results in Figure 3.6. CitNetExplorer is used for citation visualization of supply chain management in disaster response. The results depict the USA as the leader of the research, followed by China and England. Table 3.2 provides a list of sample citations related to the research keyword.

However, the topic can be narrowed down to more specific areas. The search on the topic of "supply chain management in disaster response" yields 148 publications with 272 citation links with 91 core publications in the duration of 2007–2021, as indicated in Figure 3.7, along with the sum of times cited per year.

The search "disaster supply network" yielded 253 publications on this topic from 1993 to 2019, with 3,525 articles cited. Figure 3.8 depicts the results of the search topic "blockchain technology" in the supply chain with only 58 publications in the period 2017–2019.

Interestingly, there were 4,781 journal publications with the keyword "blockchain" within the chosen search query. However, only seven are related to the disaster supply chain. The scarcity of articles forces one to look at keywords that co-occur in the published articles to illustrate the current approaches to blockchain technology applications in disaster management. The main keywords within the articles have 105

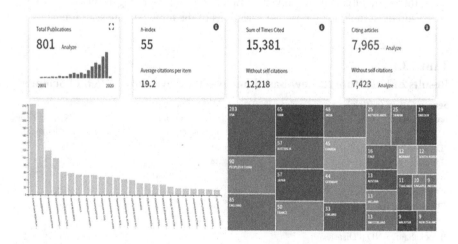

FIGURE 3.6   A tree map for disaster management along with leading research nations.

TABLE 3.2

Resulting Peer-Reviewed Journal Articles and Their Relation to Research Selected Keywords

Proponents	Research Keyword Terms		
	"Disaster Management"	"Humanitarian Aid"	"Blockchain Technology"
Balcik & Beamon, 2008		x	
Balcik et al., 2010		x	
Comfort, 2007	x		
Day et al., 2012		x	
Dubey et al., 2020			x
Holguín-Veras et al., 2012		x	
Imran et al., 2015	x		
Jahre & Jensen, 2010		x	
Kshetri, 2018; Saberi et al., 2019			x
Majchrzak et al., 2007	x		
Maon et al., 2009		x	
Papadopoulos et al., 2017		x	
Reinsberg, 2019			x
Scholten et al., 2014		x	
Stewart et al., 2009	x		
Tofighi et al., 2016		x	
Tomasini & van Wassenhove, 2009		x	
van Wassenhove, 2006		x	

links from 18 published articles. This relation is depicted in Figure 3.9—rendered from the VOSviewer tool.

The next step is to review manually review the relevant keywords. Table 3.3 presents the keywords with the most significant weight within the citation network. It is based on the results of keywords' co-occurrence network of citations using the VOSviewer tool. The keyword terms "collaboration," "humanitarian aid," and "swift trust" have the most significant weight within the citation network. Table 3.3 includes the extracted articles with the most critical keywords related to the current study.

3.5.2 RESULTS OF SYSTEMATIC LITERATURE REVIEW

The literature results suggest that blockchain applications in disaster supply chain and logistics networks are in their early stages. Moreover, we suggest that the highly cited peer-reviewed journal articles can be used to appreciate how blockchain features can address the current challenges in the field. In the next step, the most important and relevant papers from the literature are selected for manual review. Table 3.4 provides the challenges identified in the literature in the disaster supply chain along with contributions to the disaster supply network in terms of blockchain technology.

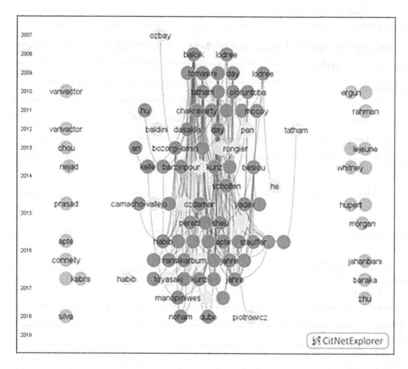

FIGURE 3.7 Visualizing supply chain management in disaster response along with the sum of citations per year.

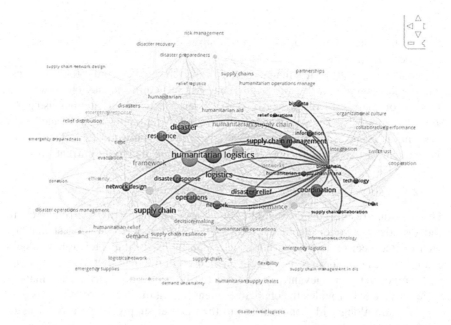

FIGURE 3.8 A citation network for disaster supply chain with blockchain links.

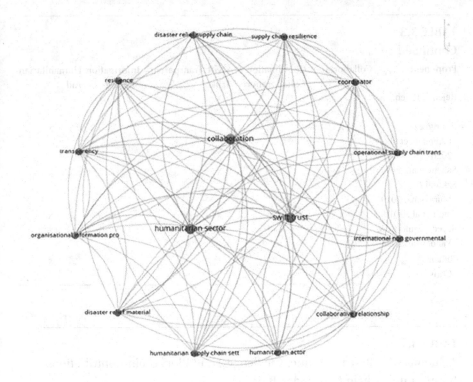

FIGURE 3.9 Blockchain in disaster management with critical keywords links.

TABLE 3.3

Peer-Reviewed Journal Articles Based on the Most Critical Research Keywords

Proponents	Collaboration	Coordination & Collaboration	Swift Trust	Transparency	Information Sharing	Humanitarian Aid
Altay et al., 2009	x		x			
Aranda et al., 2019	x		x			x
Balcik et al., 2010	x					
Dubey et al., 2020	x		x			x
Fu et al., 2020	x	x	x	x	X	x
Khan et al., 2021	x		x			x
L'Hermitte & Nair, 2021	x		x			x
Madianou, 2019				x	X	x
Ozdemir et al., 2020			x			x
Patil et al., 2020	x					x
Ransikarbum., 2015						x
Reinsberg., 2019						x

(Continued)

TABLE 3.3
Continued

Proponents	Collaboration	Coordination & Collaboration	Swift Trust	Transparency	Information Sharing	Humanitarian Aid
Rejeb & Rejeb, 2020			x			x
Rodríguez-Espíndola et al., 2020	x					x
Sahebi et al., 2017			x			x
Sakurai & Murayama, 2019						
Samir et al., 2019	x	x	x	x	X	x
Seyedsayamdost & Vanderwal, 2020			x			x
Tatham & Christopher, 2014						x

TABLE 3.4
A Summary of Research Topic, Challenges, and Blockchain Contributions to the Supply Chain Knowledge Base

Proponents	Research Topics	Challenges Identified	Blockchain Contributions
Balcik & Beamon, 2008	• Humanitarian Aid	• Lack of proper communication in relief sectors • Need for an effective and efficient network in disaster response	
Balcik et al., 2010	• Humanitarian Aid	• Lack of systematic relief chain coordination	
Day et al., 2012	• Disaster Response • Supply Chain	• Lack of demand visibility, information, coordination, trust	
Dubey et al., 2020	• Humanitarian Aid • Blockchain	• Lack of technology-enabled swift trust and transparency	A blockchain-based theoretical prototype to enhance swift trust and collaboration among relief actors
Jahre et al., 2009	• Disaster Response • Supply Chain	• Challenges in coordination and collaboration in the relief sector	
Khan et al., 2021	• Humanitarian Aid • Supply Chain	• Lack of transparency and trust, and coordination	Integration of blockchain technology and IoT to improve the humanitarian logistic performance

OL'Hermitte & Nair, 2021	• Humanitarian Aid • Supply Chain	• Lack of communication and trust	Blockchain-based platform to enable communication and share real-time information to facilitate interactions and enhance trust
Majchrzaj et al., 2007	• Communication in • Disaster Response	• Communication difficulty due to geographical dispersion • Lack of swift trust	
Maon et al., 2009	• Disaster Response • Supply Chain	• Lack of coordination in relief sectors	
Ozdemir et al., 2020; Patil et al., 2020; Sahebi et al., 2020	• Humanitarian Aid • Supply Chain	• Lack of cooperation between blockchain technology developers, donors, and aid sectors	Identify potential barriers to adopting blockchain technology in the disaster supply chain
Papadopoulos et al., 2017	• Disaster Supply • Resilience	• Importance of swift trust, information sharing, and partnerships in the supply network resiliency	
Reinsberg., 2019; Seyedsayamdost and Vanderwal (2020)	• Humanitarian • Blockchain	• Lack of systemic governance to make informed policy decisions	A blockchain-based theoretical model to enhance swift trust and collaboration among relief actors
Rodríguez-Espíndola et al., 2020	• Humanitarian Aid • Supply Chain	• Lack of accountability and poor communication	Integrate blockchain technology, IoT, and 3D printing to improve the flow of information, financial resources, and products in the humanitarian supply chain
Saberi et al., 2019	• Blockchain • Supply Chain	• Categorize barriers to blockchain adoption in the supply chain: inter-organizational, intra-organization, technical, and external	
Tomasini & Van Wassenhove, 2009	• Disaster Response • Supply Chain	• Importance of roles in humanitarian logistics	
Van Wassnhove, 2006	• Humanitarian Aid	• High level of complexity of disaster relief collaboration	

Recent technological advances have changed how we connect and distribute news and other entities. Moreover, these technologies have altered the flow of information, replaced centralized systems, and created ad hoc distributive information systems and networks. And yet, we still face inconsistency in many aspects of system management, often attributed to a lack of regular updating of information systems

(Glik, 2007). The lack of interaction and institutional overlap among involved communities generate conflicts (Schipper & Pelling, 2006). Usually, the provided aids are based on the organization's decisions, not precisely the recipient's needs.

Moreover, there is also a trend toward technologies that can handle communications challenges in humanitarian aid, with many resolutions gearing towards using computer models, artificial intelligence, and satellite imagery to enhance decision-making processes aided by the accuracy and reliability of the information. One such technology is distributed ledger technology (DLT), a consensus of replicated, shared, and synchronized digital data geographically spread across multiple sites, countries, or institutions. Unlike a centralized database, no central administrator exists (Scardovi, 2016). Noteworthy is how Coppi and Fast (2009) insist that DLT can address disaster supply chain and logistic challenges, including efficiency, scale, sustainability, and transparency. Moreover, Rajan (2018) submits that DLT can update the ledger with consensus, share transaction databases, and record the ledger with timestamped and tamper-proofed auditable history.

Finally, most DLT humanitarian support efforts focus on improving elements of cross-border transfer, donations transparency, grant management, micro-insurance, organizational governance, and reducing fraud as well as tracking support from multiple sources. In the present research, we argue for blockchain technology over DTL. Blockchain technology has a superior capability to tackle challenges associated with losses from disasters efficiently. In fact, Coppi and Fast (2009) maintain that the current projects are more limited to accountability and protection framework based on humanitarian "principles" and go on to suggest that there are limitations when it comes to humanitarian "purposes." Nevertheless, these limitations should not tarnish the viable applications of DLT. For example, in 2016, some banks tested DLTs for payments (Eyers, 2016) to see if investing in distributed ledgers is supported by their usefulness (Scardovi, 2016). In 2020, Axoni launched Veris, a DLT system that manages equity swap transactions (McDowell, 2020). A platform, which matches and reconciles post-trade data on stock swaps, is used by BlackRock Inc., Goldman Sachs Group Inc., and Citigroup, Inc. (Brush, 2021) is also based on DLT.

CONCLUSIONS

The value of theory in any field is undeniable. This chapter explores theories for disaster supply chain and logistics management, including the role of modeling and simulation and emerging technologies. To improve a situation, one needs to know and understand the situation as is. This called for a literature review to indicate what has been done in the present research context. Thus, a literature review on the supply chain in disaster management and the role of blockchain technology within disaster supply networks was conducted. Keywords were used in the Web of Science database from 1999 to 2021. The extracted data were analyzed in terms of types of methodology, domain, and indicators, with more than 600 publications meeting the criteria. However, the final analysis was conducted on less than 100 full articles. Literature also suggests that many limitations of blockchain technology still exist. Perhaps as a result of still being in the early stages, blockchain technology limitations include (Ozdemir et al., 2020; Patil et al., 2020; Sahebi et al., 2020) the following:

- Compatibility and scalability
- Cost complexities
- Data privacy, ownership, and security
- Infrastructural challenges
- Interoperability and collaboration among stakeholders
- Interorganizational complexities
- Lack of awareness and understanding among the participants
- Lack of engagement
- Legal and social frameworks
- Limited management support
- Media backlash risk
- Operational restrictions
- Technological complexities
- Value proposition uncertainty

And while there exist frameworks attempting to explain resilience in disaster supply networks, there is a scarcity of literature discussing how blockchain technology can be used to enhance the resiliency in disaster supply networks, especially those that address managing disasters and reducing lead time. Moreover, such research must be conducted while cognizant of the present limitations of the blockchain field.

REFERENCES

Adeel, A., Gogate, M., Farooq, S., Ieracitano, C., Dashtipour, K., Larijani, H., & Hussain, A. (2019). A survey on the role of wireless sensor networks and IoT in disaster management. In T. S. Durrani, W. Wang, & S. M. Forbes (Eds.), *Geological disaster monitoring based on sensor networks* (pp. 57–66). Springer. https://doi.org/10.1007/978-981-13-0992-2_5

Akram, A., & Bross, P. (2018). Trust, privacy and transparency with blockhain technology in logistics. *MCIS*. https://research.chalmers.se/publication/507357/file/507357_Fulltext.pdf

Alazawi, Z., Altowaijri, S., Mehmood, R., & Abdljabar, M. B. (2011). Intelligent disaster management system based on cloud-enabled vehicular networks. *2011 11th International Conference on ITS Telecommunications*, 361–368. https://doi.org/10.1109/ITST.2011.6060083

AliPour, F. S. (2021). *Application of a blockchain enabled model in disaster aids supply network resilience* [Dissertation, Old Dominion University Libraries]. https://doi.org/10.25777/FKR7-A212

Altay, N., & Green, W. G. (2006). OR/MS research in disaster operations management. *European Journal of Operational Research*, *175*(1), 475–493. https://doi.org/10.1016/j.ejor.2005.05.016

Altay, N., Prasad, S., & Sounderpandian, J. (2009). Strategic planning for disaster relief logistics: Lessons from supply chain management. *International Journal of Services Sciences*, *2*(2), 142–161. https://doi.org/10.1504/IJSSci.2009.024937

Anand, A., McKibbin, M., & Pichel, F. (2017, May 2). *Colored coins: Bitcoin, blockchain, and land administration*. Cadasta. 2017 World Bank Conference on Land and Poverty, Washington, DC. https://cadasta.org/resources/white-papers/bitcoin-blockchain-land/

Aqlan, F., & Lam, S. S. (2015). A fuzzy-based integrated framework for supply chain risk assessment. *International Journal of Production Economics*, *161*, 54–63. https://doi.org/10.1016/j.ijpe.2014.11.013

Aranda, D. A., Fernández, L. M. M., & Stantchev, V. (2019). Integration of Internet of Things (IoT) and Blockchain to increase humanitarian aid supply chains performance. *2019 5th International Conference on Transportation Information and Safety (ICTIS)*, 140–145. https://doi.org/10.1109/ICTIS.2019.8883757

Babich, V., & Hilary, G. (2020). OM forum—Distributed ledgers and operations: What operations management researchers should know about blockchain technology. *Manufacturing & Service Operations Management*, 22(2), 223–240. https://doi.org/10.1287/msom.2018.0752

Balcik, B., & Beamon, B. M. (2008). Facility location in humanitarian relief. *International Journal of Logistics Research and Applications*, 11(2), 101–121. https://doi.org/10.1080/13675560701561789

Balcik, B., Beamon, B. M., Krejci, C. C., Muramatsu, K. M., & Ramirez, M. (2010). Coordination in humanitarian relief chains: Practices, challenges and opportunities. *International Journal of Production Economics*, 126(1), 22–34. https://doi.org/10.1016/j.ijpe.2009.09.008

Bankes, S. (1993). Exploratory modeling for policy analysis. *Operations Research, 41*(3), 435–449. https://doi.org/10.1287/opre.41.3.435

Ben Othman, S., Zgaya, H., Dotoli, M., & Hammadi, S. (2017). An agent-based decision support system for resources' scheduling in emergency supply chains. *Control Engineering Practice, 59*, 27–43. https://doi.org/10.1016/j.conengprac.2016.11.014

Borshchev, A., & Filippov, A. (2004, July 25–29). From system dynamics and discrete event to practical agent based modeling: Reasons, techniques, tools. The 22nd International Conference of the System Dynamics Society

Brush, S. (2021, September 7). BlackRock joins blockchain platform Axoni for equity swap trades. *Bloomberg.Com*. www.bloomberg.com/news/articles/2021-09-07/blackrock-joins-blockchain-platform-axoni-for-equity-swap-trades

Calheiros, R. N., & Buyya, R. (2014). Energy-efficient scheduling of urgent bag-of-tasks applications in clouds through DVFS. *2014 IEEE 6th International Conference on Cloud Computing Technology and Science*, 342–349. https://doi.org/10.1109/CloudCom.2014.20

Cartier, A. V., Laprade, C. A., Pierri, M. V., & Worsham, D. H. (2009). *Disaster decision making: Hurricanes Katrina and Gustav in New Orleans* [Project Number: 0902, Worcester Polytechnic Institute]. https://web.wpi.edu/Pubs/E-project/Available/E-project-031609-172502/unrestricted/IQPReportFINAL.pdf

Casado-Vara, R., Prieto, J., la Prieta, F. D., & Corchado, J. M. (2018). How blockchain improves the supply chain: Case study alimentary supply chain. *Procedia Computer Science, 134*, 393–398. https://doi.org/10.1016/j.procs.2018.07.193

Cisco. (2020). *Cisco annual internet report (2018–2023): White paper* (No. c11-741490). www.cisco.com/c/en/us/solutions/collateral/executive-perspectives/annual-internet-report/white-paper-c11-741490.html

Clauson, K. A., Breeden, E. A., Davidson, C., & Mackey, T. K. (2018). Leveraging blockchain technology to enhance supply chain management in healthcare: An exploration of challenges and opportunities in the health supply chain. *Blockchain in Healthcare Today*. https://doi.org/10.30953/bhty.v1.20

Comfort, L. K. (2007). Crisis management in hindsight: Cognition, communication, coordination, and control. *Public Administration Review, 67*(s1), 189–197. https://doi.org/10.1111/j.1540-6210.2007.00827.x

Coppi, G., & Fast, L. (2009). *Blockchain and distributed ledger technologies in the humanitarian sector*. Humanitarian Policy Group. www.irisguard.com/media/4dsnvlvf/hpgblockchain.pdf

Day, J. M., Melnyk, S. A., Larson, P. D., Davis, E. W., & Whybark, D. C. (2012). Humanitarian and disaster relief supply chains: A matter of life and death. *Journal of Supply Chain Management, 48*(2), 21–36. https://doi.org/10.1111/j.1745-493X.2012.03267.x

De la Torre, L. E., Dolinskaya, I. S., & Smilowitz, K. R. (2012). Disaster relief routing: Integrating research and practice. *Socio-Economic Planning Sciences, 46*(1), 88–97. https://doi.org/10.1016/j.seps.2011.06.001

Drabek, T. E. (2007). Community processes: Coordination. In H. Rodríguez, E. L. Quarantelli, & R. R. Dynes (Eds.), *Handbook of disaster research* (pp. 217–233). Springer. https://doi.org/10.1007/978-0-387-32353-4_13

Dubey, R., Gunasekaran, A., Bryde, D. J., Dwivedi, Y. K., & Papadopoulos, T. (2020). Blockchain technology for enhancing swift-trust, collaboration and resilience within a humanitarian supply chain setting. *International Journal of Production Research, 58*(11), 3381–3398. https://doi.org/10.1080/00207543.2020.1722860

Etemadi, N., Borbon-Galvez, Y., Strozzi, F., & Etemadi, T. (2021). Supply chain disruption risk management with blockchain: A dynamic literature review. *Information, 12*(2), 70. https://doi.org/10.3390/info12020070

Eusgeld, I., Henzi, D., & Kröger, W. (2008). *Comparative evaluation of modeling and simulation techniques for interdependent critical infrastructures* (p. 50). ETH Zurich. www.bevoelkerungsschutz.admin.ch/internet/bs/de/home/themen/ski/publikationen_ski.parsys.87450.DownloadFile.tmp/comparativeevaluation.pdf

Eyers, J. (2016, November 20). Central banks look to the future of money with blockchain technology trial. *Australian Financial Review.* www.afr.com/technology/central-banks-look-to-the-future-of-money-with-blockchain-technology-trial-20161118-gss4nd

Fisher, T. (2022). 6G: What it is & when to expect it. *Lifewire.* www.lifewire.com/6g-wireless-4685524

Forrester, J. W. (1961). *Industrial dynamics.* Cambridge, MA: MIT Press.

Fu, H., Zhao, C., Cheng, C., & Ma, H. (2020). Blockchain-based agri-food supply chain management: Case study in China. *International Food and Agribusiness Management Review, 23*(5), 667–679. https://doi.org/10.22434/IFAMR2019.0152

Glik, D. C. (2007). Risk communication for public health emergencies. *Annual Review of Public Health, 28*, 33–54. https://doi.org/10.1146/annurev.publhealth.28.021406.144123

Gomes, C., Van Tendeloo, Y., Denil, J., De Meulenaere, P., & Vangheluwe, H. (2017, February 14). Hybrid system modelling and simulation with Dirac Deltas. *2017 SpringSim.* http://arxiv.org/abs/1702.04274

Gu, L., Zeng, D., Barnawi, A., Guo, S., & Stojmenovic, I. (2015). Optimal task placement with QoS constraints in geo-distributed data centers using DVFS. *IEEE Transactions on Computers, 7*(64), 2049–2059. https://doi.org/10.1109/TC.2014.2349510

Guo, H., & Yu, X. (2022). A survey on blockchain technology and its security. *Blockchain: Research and Applications, 3*(2), 100067. https://doi.org/10.1016/j.bcra.2022.100067

Haimes, Y. Y. (1981). Hierarchical holographic modeling. *IEEE Transactions on Systems, Man, and Cybernetics, 11*(9), 606–617. https://doi.org/10.1109/TSMC.1981.4308759

Hardy, M., Dodge, L., Smith, T., Vasconez, K. C., & Wunderlich, K. E. (2008). Evacuation management operations modeling assessment: Transportation modeling inventory. 15th World Congress on Intelligent Transport Systems and ITS America's 2008 Annual MeetingITS AmericaERTICOITS JapanTransCore, New York. https://trid.trb.org/view/901942

Hardy, M., Wunderlich, K., & Bunch, J. (2009). *Structuring modeling and simulation analysis for evacuation planning and operations* (FHWA—HOP-08-029). US Department of Transportation. www.semanticscholar.org/paper/Structuring-modeling-and-simulation-analysis-for-Hardy-Wunderlich/23191510450cb6ed3ee616df54e415631d37ce4f

Hassan, F. ul, Ali, A., Latif, S., Qadir, J., Kanhere, S. S., Singh, J., & Crowcroft, J. (2019). Blockchain and the future of the Internet: A comprehensive review. ArXiv: Cryptography and Security. www.scinapse.io

Hester, P. T., & Tolk, A. (2010). Applying methods of the M&S spectrum for complex systems engineering. *SCS*, 1–8.

Holguín-Veras, J., Jaller, M., Van Wassenhove, L. N., Pérez, N., & Wachtendorf, T. (2012). On the unique features of post-disaster humanitarian logistics. *Journal of Operations Management*, *30*(7–8), 494–506. https://doi.org/10.1016/j.jom.2012.08.003

Hosseini Shirvani, M., Rahmani, A. M., & Sahafi, A. (2020). A survey study on virtual machine migration and server consolidation techniques in DVFS-enabled cloud datacenter: Taxonomy and challenges. *Journal of King Saud University—Computer and Information Sciences*, *32*(3), 267–286. https://doi.org/10.1016/j.jksuci.2018.07.001

Imran, M., Castillo, C., Diaz, F., & Vieweg, S. (2015). Processing social media messages in mass emergency: A survey. *ACM Computing Surveys*, *47*(4), 67:1–67:38. https://doi.org/10.1145/2771588

Ivanov, D., Dolgui, A., Das, A., & Sokolov, B. (2019). Digital supply chain twins: Managing the ripple effect, resilience, and disruption risks by data-driven optimization, simulation, and visibility. In D. Ivanov, A. Dolgui, & B. Sokolov (Eds.), *Handbook of ripple effects in the supply chain* (pp. 309–332). Springer International Publishing. https://doi.org/10.1007/978-3-030-14302-2_15

Jahre, M., & Jensen, L. (2010). Coordination in humanitarian logistics through clusters. *International Journal of Physical Distribution & Logistics Management*, *40*(8–9), 657–674. https://doi.org/10.1108/09600031011079319

Jahre, M., Jensen, L., & Listou, T. (2009). Theory development in humanitarian logistics: A framework and three cases. *Management Research News*, *32*(11), 1008–1023. https://doi.org/10.1108/01409170910998255

Katina, P. F. (2022). Metasystem pathologies in complex system governance. In C. B. Keating, P. F. Katina, C. W. Chesterman Jr., & J. C. Pyne (Eds.), *Complex system governance: theory and practice* (pp. 241–282). Springer International Publishing. https://doi.org/10.1007/978-3-030-93852-9_9

Katina, P. F., Tolk, A., Keating, C. B., & Joiner, K. F. (2020). Modelling and simulation in complex system governance. *International Journal of System of Systems Engineering*, *10*(3), 262–292. https://doi.org/10.1504/IJSSE.2020.109739

Keating, C. B., Katina, P. F., Chesterman, C. W., & Pyne, J. C. (Eds.). (2022). *Complex system governance: Theory and practice*. Springer International Publishing. https://link.springer.com/book/10.1007/978-3-030-93852-9

Kelley, J. E., & Walker, M. R. (1959). Critical-path planning and scheduling. *Papers Presented at the December 1–3, 1959, Eastern Joint IRE-AIEE-ACM Computer Conference on—IRE-AIEE-ACM'59 (Eastern)*, 160–173. https://doi.org/10.1145/1460299.1460318

Kendall, S. (2019). *LibGuides: PubMed, web of science, or google scholar? A behind-the-scenes guide for life scientists : Which one is best: PubMed, web of science, or Google Scholar?* https://libguides.lib.msu.edu/c.php?g=96972&p=627295

Khan, M., Imtiaz, S., Parvaiz, G. S., Hussain, A., & Bae, J. (2021). Integration of Internet-of-Things with blockchain technology to enhance humanitarian logistics performance. *IEEE Access*, *9*, 25422–25436. https://doi.org/10.1109/ACCESS.2021.3054771

Kshetri, N. (2018). 1 Blockchain's roles in meeting key supply chain management objectives. *International Journal of Information Management*, *39*, 80–89. https://doi.org/10.1016/j.ijinfomgt.2017.12.005

Kushwaha, S. S., & Joshi, S. (2021). An overview of blockchain-based smart contract. In S. Smys, R. Palanisamy, Á. Rocha, & G. N. Beligiannis (Eds.), *Computer networks and inventive communication technologies* (pp. 899–906). Springer. https://doi.org/10.1007/978-981-15-9647-6_70

Lambert, D. M., Cooper, M. C., & Pagh, J. D. (1998). Supply chain management: Implementation issues and research opportunities. *The International Journal of Logistics Management*, *9*(2), 1–20. https://doi.org/10.1108/09574099810805807

L'Hermitte, C., & Nair, N.-K. C. (2021). A blockchain-enabled framework for sharing logistics resources during emergency operations. *Disasters*, *45*(3), 527–554. https://doi.org/10.1111/disa.12436

Lin, Y.-H., Batta, R., Rogerson, P. A., Blatt, A., & Flanigan, M. (2011). A logistics model for emergency supply of critical items in the aftermath of a disaster. *Socio-Economic Planning Sciences*, *45*(4), 132–145. https://doi.org/10.1016/j.seps.2011.04.003

Lohmer, J., Bugert, N., & Lasch, R. (2020). Analysis of resilience strategies and ripple effect in blockchain-coordinated supply chains: An agent-based simulation study. *International Journal of Production Economics*, *228*, 107882. https://doi.org/10.1016/j.ijpe.2020.107882

Luna, S., & Pennock, M. J. (2018). Social media applications and emergency management: A literature review and research agenda. *International Journal of Disaster Risk Reduction*, *28*, 565–577. https://doi.org/10.1016/j.ijdrr.2018.01.006

Madianou, M. (2019). The biometric assemblage: Surveillance, experimentation, profit, and the measuring of refugee bodies. *Television & New Media*, *20*(6), 581–599. https://doi.org/10.1177/1527476419857682

Majchrzak, A., Jarvenpaa, S. L., & Hollingshead, A. B. (2007). Coordinating expertise among emergent groups responding to disasters. *Organization Science*, *18*(1), 147–161. https://doi.org/10.1287/orsc.1060.0228

Manoj, B. S., & Baker, A. H. (2007). Communication challenges in emergency response. *Communications of the ACM*, *50*(3), 51–53. https://doi.org/10.1145/1226736.1226765

Maon, F., Lindgreen, A., & Vanhamme, J. (2009). Developing supply chains in disaster relief operations through cross-sector socially oriented collaborations: A theoretical model. *Supply Chain Management: An International Journal*, *14*(2), 149–164. https://doi.org/10.1108/13598540910942019

McDowell, H. (2020). Citi and Goldman Sachs go live with blockchain equity swaps platform. *The Trade*. www.thetradenews.com/citi-goldman-sachs-go-live-blockchain-equity-swaps-platform/

Mell, P., & Grance, T. (2011). *The NIST definition of cloud computing* (NIST Special Publication 800–145; pp. 1–7). National Institute of Standards and Technology. https://nvlpubs.nist.gov/nistpubs/Legacy/SP/nistspecialpublication800-145.pdf

Mijnhardt, W. (2007, June 21). HISTCITE™; Bibiliographic analysis and visualization software. *Research-Management In Management-Research [RMIMR]*. https://rmimr.wordpress.com/2007/06/21/histcite%e2%84%a2-bibiliographic-analysis-and-visualization-software/

Montazerolghaem, A., Yaghmaee, M. H., & Leon-Garcia, A. (2020). Green cloud multimedia networking: NFV/SDN based energy-efficient resource allocation. *IEEE Transactions on Green Communications and Networking*, *4*(3), 873–889. https://doi.org/10.1109/TGCN.2020.2982821

Mustapha, K., Mcheick, H., & Mellouli, S. (2013). Modeling and simulation agent-based of natural disaster complex systems. *Procedia Computer Science*, *21*, 148–155. https://doi.org/10.1016/j.procs.2013.09.021

NSF. (2006). *Simulation-based engineering science: Revolutionazing engineering science through simulation* (pp. 1–66). National Science Foundation. www.nsf.gov/pubs/reports/sbes_final_report.pdf

Ozdemir, A. I., Erol, I., Ar, I. M., Peker, I., Asgary, A., Medeni, T. D., & Medeni, I. T. (2020). The role of blockchain in reducing the impact of barriers to humanitarian supply chain management. *The International Journal of Logistics Management*, *32*(2), 454–478. https://doi.org/10.1108/IJLM-01-2020-0058

Papadopoulos, T., Gunasekaran, A., Dubey, R., Altay, N., Childe, S. J., & Fosso-Wamba, S. (2017). The role of Big Data in explaining disaster resilience in supply chains for sustainability. *Journal of Cleaner Production*, *142*, 1108–1118. https://doi.org/10.1016/j.jclepro.2016.03.059

Patil, A., Shardeo, V., Dwivedi, A., & Madaan, J. (2020). An integrated approach to model the blockchain implementation barriers in humanitarian supply chain. *Journal of Global Operations and Strategic Sourcing*, *14*(1), 81–103. https://doi.org/10.1108/JGOSS-07-2020-0042

Paton, D., & Flin, R. (1999). Disaster stress: An emergency management perspective. *Disaster Prevention and Management*, *8*(4), 261–267. https://doi.org/10.1108/09653569910283897

Pidd, M., de Silva, F. N., & Eglese, R. W. (1996). A simulation model for emergency evacuation. *European Journal of Operational Research*, *90*(3), 413–419.

Plageras, A. P., Psannis, K. E., Stergiou, C., Wang, H., & Gupta, B. B. (2018). Efficient IoT-based sensor BIG Data collection—processing and analysis in smart buildings. *Future Generation Computer Systems*, *82*, 349–357. https://doi.org/10.1016/j.future.2017.09.082

Queiroz, M. M., Telles, R., & Bonilla, S. H. (2019). Blockchain and supply chain management integration: A systematic review of the literature. *Supply Chain Management: An International Journal*, *25*(2), 241–254. https://doi.org/10.1108/SCM-03-2018-0143

Rajan, S. G. (2018). *Analysis and design of systems utilizing blockchain technology to accelerate the humanitarian actions in the event of natural disasters* [Thesis, Massachusetts Institute of Technology]. https://dspace.mit.edu/handle/1721.1/118526

Ransikarbum, K. (2015). *Disaster management cycle-based integrated humanitarian supply network management* [Dissertation, Clemson University].

Ray, P. P. (2018). An introduction to dew computing: Definition, concept and implications. *IEEE Access*, *6*, 723–737. https://doi.org/10.1109/ACCESS.2017.2775042

Reinsberg, B. (2019). Blockchain technology and the governance of foreign aid. *Journal of Institutional Economics*, *15*(3), 413–429. https://doi.org/10.1017/S1744137418000462

Rejeb, A., & Rejeb, K. (2020). Blockchain and supply chain sustainability. *Logforum*, *16*(3), 363–372. https://doi.org/10.17270/J.LOG.2020.467

Richey, R. G. (2009). The supply chain crisis and disaster pyramid: A theoretical framework for understanding preparedness and recovery. *International Journal of Physical Distribution & Logistics Management*, *39*(7), 619–628. https://doi.org/10.1108/09600030910996288

Rinaldi, S. M., Peerenboom, J., & Kelly, T. K. (2001). Identifying, understanding, and analyzing critical infrastructure interdependencies. *IEEE Control Systems*, *21*(6), 11–25. https://doi.org/10.1109/37.969131

Risius, M., & Spohrer, K. (2017). A blockchain research framework. *Business & Information Systems Engineering*, *59*(6), 385–409. https://doi.org/10.1007/s12599-017-0506-0

Rodrigo, N., Perera, S., Senaratne, S., & Jin, X. (2018). Blockchain for construction supply chains: A literature synthesis. Proceedings of the 11th International Cost Engineering Council (ICEC) World Congress & the 22nd Annual Pacific Association of Quantity Surveyors Conference. https://researchdirect.westernsydney.edu.au/islandora/object/uws%3A49027/

Rodríguez-Espíndola, O., Chowdhury, S., Beltagui, A., & Albores, P. (2020). The potential of emergent disruptive technologies for humanitarian supply chains: The integration of blockchain, Artificial Intelligence and 3D printing. *International Journal of Production Research*, *58*(15), 4610–4630. https://doi.org/10.1080/00207543.2020.1761565

Saberi, S., Kouhizadeh, M., Sarkis, J., & Shen, L. (2019). Blockchain technology and its relationships to sustainable supply chain management. *International Journal of Production Research*, *57*(7), 2117–2135. https://doi.org/10.1080/00207543.2018.1533261

Sahebi, I. G., Arab, A., & Sadeghi Moghadam, M. R. (2017). Analyzing the barriers to humanitarian supply chain management: A case study of the Tehran Red Crescent Societies. *International Journal of Disaster Risk Reduction*, *24*, 232–241. https://doi.org/10.1016/j.ijdrr.2017.05.017

Sahebi, I. G., Masoomi, B., & Ghorbani, S. (2020). Expert oriented approach for analyzing the blockchain adoption barriers in humanitarian supply chain. *Technology in Society*, *63*, 101427. https://doi.org/10.1016/j.techsoc.2020.101427

Sakurai, M., & Murayama, Y. (2019). Information technologies and disaster management: Benefits and issues. *Progress in Disaster Science*, *2*, 100012. https://doi.org/10.1016/j.pdisas.2019.100012

Samir, E., Azab, M., & Jung, Y. (2019). Blockchain guided trustworthy interactions for distributed disaster management. *2019 IEEE 10th Annual Information Technology, Electronics and Mobile Communication Conference (IEMCON)*, 0241–0245. https://doi.org/10.1109/IEMCON.2019.8936147

Scardovi, C. (2016). *Restructuring and innovation in banking*. Springer.

Schipper, L., & Pelling, M. (2006). Disaster risk, climate change and international development: Scope for, and challenges to, integration. *Disasters, 30*(1), 19–38. https://doi.org/10.1111/j.1467-9523.2006.00304.x

Schneider, K., Chen-Ching, L., & Paul, J. P. (2006). Assessment of interactions between power and telecommunications infrastructures. *Power Systems, IEEE Transactions On, 21*(3), 1123–1130. https://doi.org/10.1109/tpwrs.2006.876692

Scholten, K., Sharkey Scott, P., & Fynes, B. (2014). Mitigation processes: Antecedents for building supply chain resilience. *Supply Chain Management: An International Journal, 19*(2), 211–228. https://doi.org/10.1108/SCM-06-2013-0191

Sebatli, A., Cavdur, F., & Kose-Kucuk, M. (2017). Determination of relief supplies demands and allocation of temporary disaster response facilities. *Transportation Research Procedia, 22*, 245–254. https://doi.org/10.1016/j.trpro.2017.03.031

Sementelli, A. (2007). Toward a taxonomy of disaster and crisis theories. *Administrative Theory & Praxis, 29*(4), 497–512. https://doi.org/10.1080/10841806.2007.11029615

Seyedsayamdost, E., & Vanderwal, P. (2020). From good governance to governance for good: Blockchain for social impact. *Journal of International Development, 32*(6), 943–960. https://doi.org/10.1002/jid.3485

Sheldon, R. (2021). A timeline and history of blockchain technology [Tech Accelerator]. *WhatIs.Com.* www.techtarget.com/whatis/feature/A-timeline-and-history-of-blockchain-technology

Stewart, G. T., Kolluru, R., & Smith, M. (2009). Leveraging public-private partnerships to improve community resilience in times of disaster. *International Journal of Physical Distribution & Logistics Management, 39*(5), 343–364. https://doi.org/10.1108/09600030910973724

Swaminathan, J. M., Smith, S. F., & Sadeh, N. M. (1998). Modeling supply chain dynamics: A multiagent approach*. *Decision Sciences, 29*(3), 607–632. https://doi.org/10.1111/j.1540-5915.1998.tb01356.x

Syafrudin, M., Alfian, G., Fitriyani, N. L., & Rhee, J. (2018). Performance analysis of IoT-based sensor, big data processing, and machine learning model for real-time monitoring system in automotive manufacturing. *Sensors, 18*(9), 2946. https://doi.org/10.3390/s18092946

Tabaklar, T., Halldórsson, Á., Kovács, G., & Spens, K. (2015). Borrowing theories in humanitarian supply chain management. *Journal of Humanitarian Logistics and Supply Chain Management, 5*(3), 281–299. https://doi.org/10.1108/JHLSCM-07-2015-0029

Tatham, P., & Christopher, M. (Eds.). (2014). *Humanitarian logistics: Meeting the challenge of preparing for and responding to disasters*. Kogan Page.

The Writing Center. (2022). Literature reviews. *The Writing Center.* https://writingcenter.unc.edu/tips-and-tools/literature-reviews/

Tian, F. (2017). A supply chain traceability system for food safety based on HACCP, blockchain amp; Internet of things. *2017 International Conference on Service Systems and Service Management*, 1–6. https://doi.org/10.1109/ICSSSM.2017.7996119

Tofighi, S., Torabi, S. A., & Mansouri, S. A. (2016). Humanitarian logistics network design under mixed uncertainty. *European Journal of Operational Research, 250*(1), 239–250. https://doi.org/10.1016/j.ejor.2015.08.059

Tomasini, R. M., & Van Wassenhove, L. N. (2009). From preparedness to partnerships: Case study research on humanitarian logistics. *International Transactions in Operational Research, 16*(5), 549–559. https://doi.org/10.1111/j.1475-3995.2009.00697.x

Townsend, F. F. (2006). *The federal response to Hurricane Katrina: Lessons learned.* US Government Printing Office.

US DHS. (2016). *Strategic principles for securing the internet of things (IoT)* (Version 1.0). US Department of Homeland Security. www.dhs.gov/news/2016/11/15/dhs-releases-strategic-principles-securing-internet-things

US DHS. (2022). *Feature Article: S&T Report Peers into the Future of 5G & 6G.* www.dhs.gov/science-and-technology/news/2022/03/24/feature-article-st-report-peers-future-5g-6g

van Eck, N. J., & Waltman, L. (2014). CitNetExplorer: A new software tool for analyzing and visualizing citation networks. *Journal of Informetrics, 8*(4), 802–823. https://doi.org/10.1016/j.joi.2014.07.006

Van Wassenhove, L. N. (2006). Humanitarian aid logistics: Supply chain management in high gear. *Journal of the Operational Research Society, 57*(5), 475–489. https://doi.org/10.1057/palgrave.jors.2602125

Verma, D. K., Katheria, V., & Khaliq, M. (2019). Use cases and applications of blockchain technology in IT industry. *International Journal of Computer Sciences and Engineering, 7*(4), 716–720. https://doi.org/10.26438/ijcse/v7i4.716720

Weber, I., Lu, Q., Tran, A. B., Deshmukh, A., Gorski, M., & Strazds, M. (2019). A platform architecture for multi-tenant blockchain-based systems. *2019 IEEE International Conference on Software Architecture (ICSA)*, 101–110. https://doi.org/10.1109/ICSA.2019.00019

Xu, R., Zhang, L., Zhao, H., & Peng, Y. (2017). Design of network media's digital rights management scheme based on blockchain technology. *2017 IEEE 13th International Symposium on Autonomous Decentralized System (ISADS)*, 128–133. https://doi.org/10.1109/ISADS.2017.21

Yoo, M., & Won, Y. (2018). A study on the transparent price tracing system in supply chain management based on blockchain. *Sustainability, 10*(11), 4037. https://doi.org/10.3390/su10114037

Zīle, K., & Strazdiņa, R. (2018). Blockchain use cases and their feasibility. *Applied Computer Systems, 23*(1), 12–20. https://doi.org/10.2478/acss-2018-0002

Ziolkowski, R., Miscione, G., & Schwabe, G. (2018, December 16). Consensus through blockchains: Exploring governance across inter-organizational settings. In: *International Conference of Information Systems (ICIS 2018)*. ICIS. https://doi.org/10.5167/uzh-160378

Zou, N., Yeh, S.-T., Chang, G.-L., Marquess, A., & Zezeski, M. (2005). Simulation-based emergency evacuation system for Ocean City, Maryland, during hurricanes. *Transportation Research Record, 1922*(1), 138–148. https://doi.org/10.1177/0361198105192200118

4 Towards Blockchain Models in Disaster Management

4.1 RESEARCH DESIGN

Research should be guided by the study's focus and data analysis (Whyte, 1984). The same logic is applied in the present research in the search for an enhanced framework for disaster supply chain and logistics management. First, the core agents of disaster management networks need to be defined. The aim is to use a blockchain platform for zero-confirmation transactions using smart contracts-enabled simulation in hyperconnected logistics. The simulation can play an essential role in demand provision in the network. DVFS algorithm is used to synchronize the combination of blockchain technology and IoT. This approach is necessary to enable tracking within the model.

The overall strategy enables the integration of the different components of the study coherently and logically, thereby ensuring the research blueprint, including data collection, measurement, and data analysis. Figure 4.1 depicts the overall research design. This study is based on a systematic literature review that articulated challenges and shaped the goal of the research. Subsequently, a model was developed and simulated. Data is generated using the simulation of several situations. The analysis is performed to develop responses to the research questions. The model is "validated" compared to recent credible and published findings. The last step is reporting the discoveries of the research.

Instruments are required for the emergency teams to make efficient interventions. And in the present case, we suggest "in vitro" simulations since they are relatively risk-free and can provide insights into the behavior of the situation via several significant iterations. This study uses a methodical approach to model disaster supply chain and communication networks. The simulated analysis includes impacts of the proposed model on disaster management resilience and identification of possible solutions to optimize supply chain disaster aids management.

4.2 A MODEL FOR THE DISASTER SUPPLY CHAIN

Obviously, the development of a disaster supply chain and logistics is not an easy feat. First, a robust understanding of interconnectedness is required. Second, the development and management of such a network require understanding the parameters of the entire system of interest and its dynamics. Richey (2009) notes that a critical aspect is understanding that collaboration is the glue that holds organizations together in disasters. In a disaster, different organizations may pursue several different and, in some cases, conflicting objectives. With the aforementioned in mind, the

DOI: 10.1201/9781003336082-4

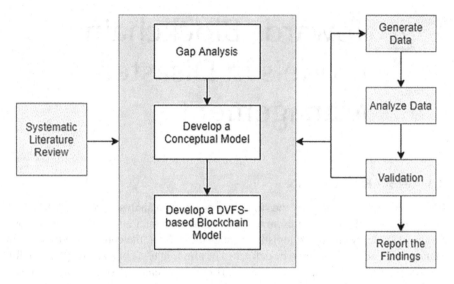

FIGURE 4.1   The proposed research design.

present research emphasizes a decision management framework for a disaster relief supply chain and logistics that considers the complexities of disasters. The developed model has a basis in limitations of current literature, is grounded in established research, and uses simulated data streams.

Moreover, another critical challenge in the disaster supply chain is the need for quick and effective distribution of items to the affected population (Thomas & Kopczak, 2005). Therefore, the present research also includes aspects of prioritizing the needs. In this case, there is a need to address non-essential goods and ensure that participants enable the swift trust to provide collaboration channels.

Despite the complexity, information technology can support each phase of disaster management by providing real-time monitoring, online dashboards, interactive communication, and collaboration (Imran et al., 2015). An often ignored approach (perhaps because current society is always seeking fame and power) for quick distribution is *decentralization*. Moreover, the technology is readily available. For example, DLT has been applied in disaster management to enhance efficiency and sustainability as well as transparency (Coppi & Fast, 2009). Data are also required to serve and monitor the disaster network. These data centers need to be independent in terms of hardware, infrastructure, and implementation and yet have the ability to share resources. Users can then access the shared resources via cloud computing.

Cloud computing is one of the resolutions that can be used to provide service in a wide range of scenarios and at a low cost (Javadpour et al., 2018). Since the disaster supply networks need to have distributed computing and services, virtual machines can enhance the performance without interruption in transferring to other nodes and dynamically change the resource amounts allocated to a client. Moreover, cloud data centers can be used to reduce operational costs and improve service quality. Furthermore, Javadpour et al. (2018) posit that the application of the load balancing technique can also help optimize the load distribution among various stages and eliminate overloading processing on one of the hosts. DVFS can be used for the robust blockchain-based decentralized resource management framework to reduce energy costs. The following section describes how a blockchain-based disaster management model can employ a DVFS algorithm to prevent overloading and reduce energy consumption and time of whole system processing.

4.3   BLOCKCHAIN-ENABLED PROTOTYPING

A critical review of several recent disasters suggests a need for network authority in organizing disaster relief supply chains (Kumar & Havey, 2013). Such efforts would involve the development of robust communication plans as well as coordinating among efforts and responses. Moreover, all involved parties would need to be offered an opportunity to be responsible for their domain expertise during the relief operations. Such a network could also involve industry, government agencies, military, and nongovernmental entities to manage better and coordinate relief efforts. Constructing a reliable disaster management system that can be practical is a complex process. It has multiple challenges, including the extraction of demand requests of the victims, establishing a framework to address time-sensitive needs, and allocating decision-making. New data-driven methods have the potential to tackle such challenges, especially since modern technological advances can enable multi-directional communication among parties and provide fashionable means of interfacing (Schempp et al., 2019).

In the proposed prototype, the first portion includes a set of transactions (i.e., C1, C2, C3) that are applied from a set of clients (AliPour, 2021). In this case, the clients are the IoT nodes representing the enforced flows in a disaster. The nodes, along with some servers, are included in a network. The round-robin (RR) algorithm on virtual machines in the IoT format is randomly executed and assigned to the virtual machines. Algorithms are used to calculate how to manage the resources and supplies. Embedding the DVFS algorithm enables online data streams to be sorted and allocated to the readiest server to minimize the failures of the responsibilities.

A matrix is created that can generate flows in various conditions and simulate realistic disaster conditions. The matrix considers the supply chain process for the set of clients. The model aims to allocate inputs of clients in the supply chain and manage the disaster data in the least time possible with high implementation as a means to enhance efficiency. In this case, tasks are considered as a high load of stream data entering the network. The model aims to handle those tasks the utmost efficiently. The tasks will be added to blocks and pass the chain stage. The highlight of the model is to determine which server in which virtual machine is the best to run the chain and how the virtual machines should be sorted (AliPour, 2021).

The second segment expresses the blockchain calculation mechanism within the various chains where DVFS is applied to analyze and sort the waiting queue of processes. Different physical and virtual machines are used within the queue to update the calculations. DVFS is used to identify the location within the network where the disaster occurred due to the huge number of requests from the clients in the transaction section. Therefore, using DVFS provides a queueing algorithm that can

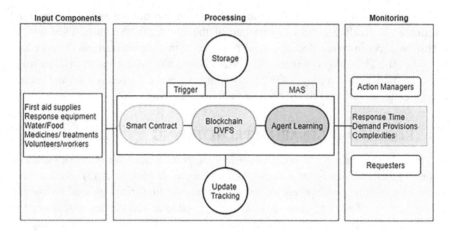

FIGURE 4.2 The proposed three-phase conceptual blockchain model for disaster management.

be processed, determines where the requirements are stored, and monitors them. Hence, the whole network is transparent and can be monitored by anyone within the network. This tracking and monitoring results are represented in the different variables: system delays, energy conception, system errors, number of actions, successful immigration, and system throughput.

Numerous components are considered inputs (and updating) for the network. The data are stored in the storage component activated with a trigger and reviewed by the agents. Binary agent-based conditions are used to consider the relationship between the blockchains and the DVFS. Smart contracts are sorted using DVFS and then analyzed. When the agent is activated, a relationship is considered between the update tracking and action management. Multi-agent systems (MAS) enable interactions among the entities within the network. Figure 4.2 depicts the three phases of the suggested conceptual model. The summation of these three steps would be the disaster management stage which can monitor the affected area and the response team.

Generating encrypted information through cryptography is critical to this process. It is important to recall that the science of using mathematical rules for the cryptography of known data is based on encryption. When the safe node identifies the shortest path, the public keys of the middle layer nodes are collected via blockchain encryption. An encryption and blockchain technology package using private keys for all the middle nodes is applied. The control stage is based on the agreements set on the server. In this control stage, transactions are reviewed with DFVS and a specific space is allocated for each of them and added as a block to the chain. Based on Pérez-Solà et al. (2019), this method includes two phases: queue development and the phase of coding and mining. This would be based on packages labeled "Hello" to the distributed transactions but waited to be added to the blockchain. The indexes intended and used in the equations are shown as follows (AliPour, 2021):

e is the rate of stream entrance to the network
e_k is the rate of external transaction entrance
Y_k is the number of requests to clients

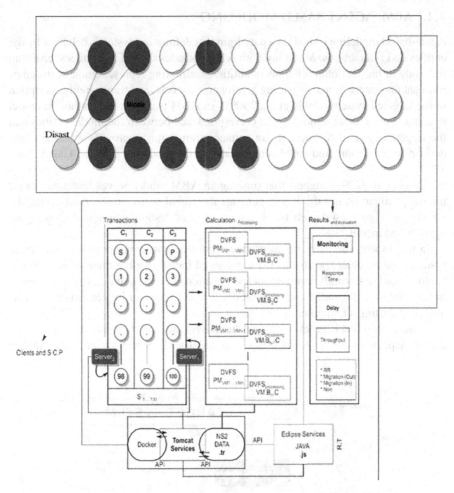

FIGURE 4.3 A demonstrated blockchain-enabled DVFS-based disaster management model execution.

i is the confirmation request rate (number of confirmations)
t is the initial waiting time
C_{ai} is the square index for the external variety
C_{si} is the square index for the internal variety
P_i is the work traffic coefficient
m_i is the required queue time
W is the weight
Total W_i is the total weights applied to the network
Pq_i is the probability of waiting in the queue
N_i is the number of transactions in the queue

Figure 4.3 depicts a demonstrated execution of the proposed blockchain-enabled DVFS-based disaster management model.

4.4 ABM: AGENT-BASED MODELING

Agent-based modeling (ABM) is a paradigm that defines the system's behavior by the entities and interactions. Modeling with ABM includes several advantages: enabling the study of bidirectional relations of entities, facilitating representation of the environment and interactions, providing heterogenic models, and transparent description of the targeted system (Galán et al., 2009). An ABM is a practical method to model systems that facilitate a more direct correspondence between the model entities and the targeted system. As such, it can enhance transparency, provide soundness and descriptive precision, and enable consistency of the modeling process (Galán et al., 2009).

Galán et al. (2009) suggest that running an ABM model proves that a particular micro-specification is sufficient to generate the global behavior observed during the simulation. Figure 4.4 depicts the different stages of designing and implementing an agent-based model.

In this model, the thematician produces the first conceptualization of the target system. The modeler transforms the non-formal model that the thematician aims to explore into the (formal) requirement specifications that the computer scientist needs to formulate the (formal) executable model. The programmer implements an executable model using a computer.

In the end, ABM is practical where the individuals and their interactions are the critical aspects of the system (Collins et al., 2020). MAS offers a natural metaphor

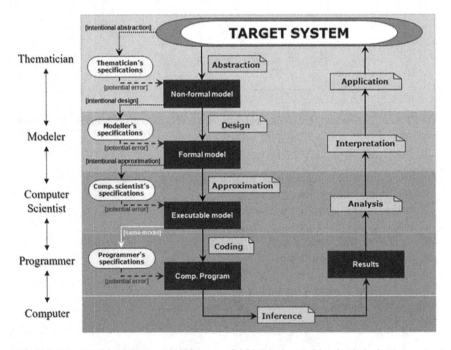

FIGURE 4.4   Different phases and elements of ABM design and implementation.

for meta-scheduling function implementation. Agents cooperate to improve the performance of the entire system. MAS is an effective tool applicable for cases where a large number of dynamic interacting entities should be modeled by modeling the collaboration among the teams of agents (Buford et al., 2006). The agents use their knowledge to make decisions and perform actions on the environment to solve the allocated tasks. The MAS is practical for cloud computer networks (Dorri et al., 2018). The main features of MAS are efficiency, low cost, flexibility, and reliability. In MAS, the agents and their relations are modeled using graphs, where each vertex represents an agent, and the edge between two agents indicates the communication. The tasks are allocated to autonomous entities (agents). Each agent decides on an appropriate activity to resolve the task based on the aim of the system while still considering action history, interactions, and multiple inputs (Dorri et al., 2018).

CONCLUSIONS

Evoking modeling and simulation draw attention to physical experimentation in which computers are used to calculate the results of some physical phenomenon. Moreover, a computer can be used to build a mathematical model that contains all the parameters of a physical phenomenon and represent that physical model in virtual form. This permits conditions to be applied, which allows experiments for physical models through that simulation (Katina et al., 2020). This approach is efficient as it voids a need for actual physical experimentation, which can be costly and time-consuming, as in disaster management. Beyond these benefits, M&S facilitates understanding a system's behavior without actually testing the system in the real world and can be used for training purposes, decision support, and a greater understanding of complex relationships.

Disaster supply chains are complicated networks, so a robust understanding of interconnectedness is required. Despite the complexity, information technologies (e.g., DTL) can be used to support each phase of disaster supply management by providing real-time monitoring, online dashboards, interactive communication, and collaboration. This chapter shows how a blockchain-enabled DVFS-based model can be developed in a decentralized matter strengthened by agent-based modeling to affect critical issues in disaster supply chain operations: cost reduction, decentralized services, increased transparency, and improved efficiency.

REFERENCES

AliPour, F. S. (2021). *Application of a blockchain enabled model in disaster aids supply network resilience* [Dissertation, Old Dominion University Libraries]. https://doi.org/10.25777/FKR7-A212

Buford, J. F., Jakobson, G., & Lewis, L. (2006). Multi-agent situation management for supporting large-scale disaster relief operations. *International Journal of Intelligent Control and Systems*, *11*(4), 284–295.

Collins, A. J., Etemadidavan, S., & Pazos-Lago, P. (2020). A human experiment using a hybrid agent-based model. 2020 Winter Simulation Conference (WSC), 1016–1026. https://doi.org/10.1109/WSC48552.2020.9384113

Coppi, G., & Fast, L. (2009). *Blockchain and distributed ledger technologies in the humanitarian sector*. Humanitarian Policy Group. www.irisguard.com/media/4dsnvlvf/hpgblockchain.pdf

Dorri, A., Kanhere, S. S., & Jurdak, R. (2018). Multi-agent systems: A survey. *IEEE Access, 6*, 28573–28593. https://doi.org/10.1109/ACCESS.2018.2831228

Galán, J. M., Izquierdo, L., Izquierdo, S., Santos, J., Olmo, R. del, Lopez-Paredes, A., & Edmonds, B. (2009). Errors and artefacts in agent-based modelling. *Journal of Artificial Societies and Social Simulation, 12*(1), 1–19.

Imran, M., Castillo, C., Diaz, F., & Vieweg, S. (2015). Processing social media messages in mass emergency: A survey. *ACM Computing Surveys, 47*(4), 67:1–67:38. https://doi.org/10.1145/2771588

Javadpour, A., Wang, G., Rezaei Badafshani, S., & Chen, S. (2018). Power curtailment in cloud environment utilising load balancing machine allocation. *The 15th IEEE International Conference on Ubiquitous Intelligence and Computing (UIC 2018)*, 1364–1370. https://doi.org/10.1109/SmartWorld.2018.00237

Katina, P. F., Tolk, A., Keating, C. B., & Joiner, K. F. (2020). Modelling and simulation in complex system governance. *International Journal of System of Systems Engineering, 10*(3), 262–292. https://doi.org/10.1504/IJSSE.2020.109739

Kumar, S., & Havey, T. (2013). Before and after disaster strikes: A relief supply chain decision support framework. *International Journal of Production Economics, 145*(2), 613–629.

Pérez-Solà, C., Delgado-Segura, S., Navarro-Arribas, G., & Herrera-Joancomartí, J. (2019). Double-spending prevention for Bitcoin zero-confirmation transactions. *International Journal of Information Security, 18*(4), 451–463. https://doi.org/10.1007/s10207-018-0422-4

Richey, R. G. (2009). The supply chain crisis and disaster pyramid: A theoretical framework for understanding preparedness and recovery. *International Journal of Physical Distribution & Logistics Management, 39*(7), 619–628. https://doi.org/10.1108/09600030910996288

Schempp, T., Zhang, H., Schmidt, A., Hong, M., & Akerkar, R. (2019). A framework to integrate social media and authoritative data for disaster relief detection and distribution optimization. *International Journal of Disaster Risk Reduction, 39*, 101143. https://doi.org/10.1016/j.ijdrr.2019.101143

Thomas, A., & Kopczak, L. R. (2005). From logistics to supply chain management: The path forward in the humanitarian sector. *Fritz Institute, 15*, 1–15.

Whyte, W. F., & Whyte, K. K. (1984). *Learning from the field: A guide from experience*. Sage Publications.

5 Results of Blockchain-Enabled Simulation

5.1 MODEL PARAMETERS

Iteration is the repetition of a process to generate a (possibly unbounded) sequence of outcomes. Each process repetition is a single iteration, and each iteration's outcome is the starting point of the next iteration. In mathematics and computer science, iteration (along with the related technique of recursion) is a standard element of algorithms. The results of 100 iterations simulating the DVFS model using a data stream are presented. Figure 5.1 depicts the simulation structure and tools used to represent the model's components.

A list of items considered critical requirements in case of a disaster is created to be used as input to the system (AliPour, 2021). This model uses Liberatore et al.'s (2013) view of the most critical requirements in disaster response, as indicated in Figure 5.2.

The data is randomly generated using a normal distribution function with a mean of 40,000 and a standard deviation of 10,000, as indicated:

Index: I
Million Number of Tasks (Estimation of The Processing Time for Each Task in Cloudlet): A random number of normal distribution functions with a mean of 40,000 and a standard deviation of 10,000
Input Data Size: 3000 KB
Output Data Size: 3000 KB
Input Time: A random number of inverses of Poisson aggregation distribution function with value $\lambda = 100$

5.2 DESCRIPTION OF THE MODEL

Modeling and simulation is the use of a physical or logical representation of a given system to generate data and help determine decisions or make predictions about the system. As a process of conducting an abstraction of a system for a particular aim, computational modeling is a formal representation animated by the computer to generate the model's outcome via a computer program, algorithm, or equations. Moreover, computer simulations can also be seen as interference tools that can go beyond mathematical tractability. In the present case, an algorithm was developed as follows (AliPour, 2021):

Start
Entrance (Disaster with Stream Data)

DOI: 10.1201/9781003336082-5

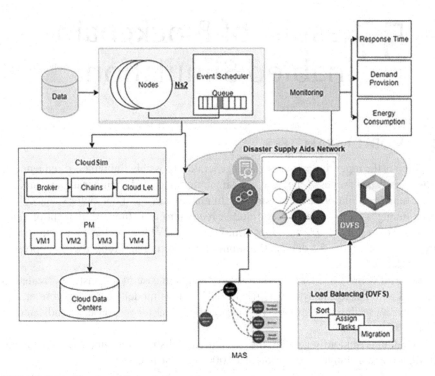

FIGURE 5.1   Structure of simulation and components.

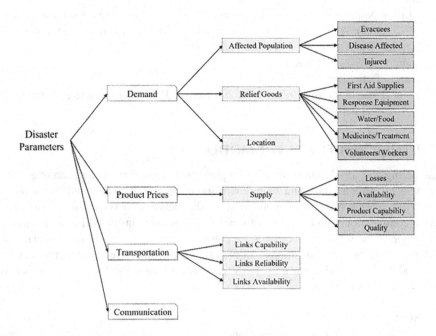

FIGURE 5.2   Essential requirements in disaster response operations.

Check: The receiver
Check: Transaction monitoring
Check: Receive data in DVFS

Run: Distributed transactions

$$e = \sum_{k}^{1} e_k$$

$$Y = \sum_{k}^{1} Y_k e_k$$

represents a crisis in level K with the entrance rate variant of $i = \{1,2,3, \ldots\}$ $T = \Sigma_i^1 e_i$ represents waiting time for monitoring transactions in each workstation if C_{ai} (square index for external variety) and C_{si} (square index for the internal variety, then

P_i is the work traffic coefficient
m_i is the required queue time

$$Total\ W_i = \left(\frac{C_{ai} + C_{si}}{2} \right) \left(\frac{p_i^{\sqrt{2(m_i + 1)} - 1}}{m_i(1 - p_i)} \right)$$

$$p_{qi} = p(N_i > m_i) = \frac{(m_i p_i)^{m_i}}{m_i!(1 - p_i)} \left(\frac{(m_i p_i)^{m_i}}{m_i!(1 - p_i)} + \sum_{j=0}^{m_i-1} \left(\frac{(m_i p_i)^i}{j!} \right) - 1 \right)$$

If ===> "entry rate is bigger than foreign transaction" > 0

Yes: Calculate the required queue time
Then:
Represents Agent-based
No: CAL
Number of transactions in the queue

Run: Trigger for Agents (Selections)

CheckOutData (Disaster with Stream data)
Check: The receiver
Check: Transaction monitoring
Check: CheckOutData data in DVFS
Entrance (Disaster with Stream Data) Check: The receiver
Check: Transaction monitoring Check: Receive data in DVFS

End

In the function, λ is the input processing time. The processing time is a stream type in the network and can be used to express space and time complexity in virtual environments. We also assume a normal distribution for the input data. This allows for outputs to be plotted in a range of (0–1). The applied function to the simulation is based on Equations 1 and 2:

$$f(x;\mu,\sigma^2) = \frac{1}{\sqrt{2\pi\sigma^2}} e^{-\left(\frac{x-\pi}{2\sigma^2}\right)}, \pi \in \mathbb{R} \tag{5.1}$$

$$f(k;\lambda) = \frac{\lambda^k e^{-\lambda}}{k!} \tag{5.2}$$

Figure 5.3 depicts the flowchart of how to execute the proposed theoretical model. Obviously, it must be stated that access to reliable information in the disaster supply chain and logistics is essential in such a model as it affects accuracy. In such a simulation, big data is generated. Big data contains greater variety, arriving in increasing volumes and with more velocity. And in such cases, one needs tools for making sense of the data (Iglesias et al., 2020). For the data generated during a disaster by the active participants, streamed big data processing frameworks are required to support disaster management.

Microprocessor without Interlocked Pipeline Stages (MIPS) is applied for analyzing and measuring the connected device's processing power. Originally, MIPS investigated a type of instruction set architecture (ISA), now called reduced instruction set computer (RISC), its implementation as a microprocessor with very large scale integration (VLSI) semiconductor technology, and the effective exploitation of RISC architectures with optimizing compilers. For the current research, datasets are added using BChainCloudSim, BigData, and CloudSim libraries. In the first step of the model, the CloudSim library is introduced, including libraries for the cores and entities. The model regarding the chain's policies is provided within this library which is based on First In First Out (FIFO). FIFO is used as a basis for recording all inputs into the system. The Cloudlet (a subset of the blockchain library) is a cloud data center supporting devices' interactive applications with less delay. Cloudlet is used based on various information represented with UserId and the required RelocationId for the allocated tasks in each virtual machine to consider and record the BChain data in the data center. To do the simulation, a Docker model is required to store the data within it. Application Programming Interface (API) links the network code and Docker. A Tomcat service is needed to consider the web-based API. Tomcat provides a "pure Java" HTTP web server environment in which Java code can run. Besides, Java enables the execution of the requests (tasks) in a blockchain structure.

5.3   DESIGN OF THE SIMULATION STUDY

Ns2 simulator is used in this analysis. Ns is a discrete event simulator targeted at networking research. Ns provides substantial support for the simulation of TCP, routing, and multicast protocols over wired and wireless (local and satellite) networks. It is an open-source tool that contains a blockchain library. Moreover, Ns2 can test complex

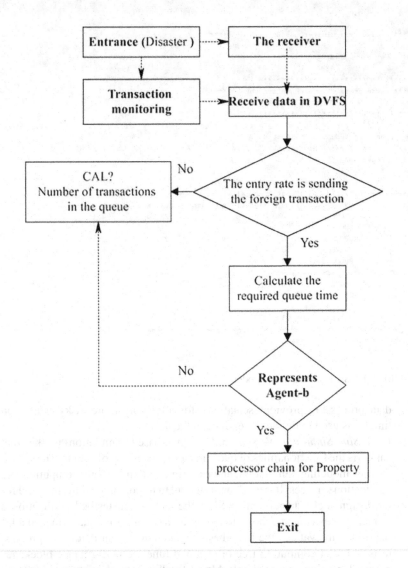

FIGURE 5.3   The flowchart of model execution.

scenarios, and results can be obtained quickly, enabling further testing of ideas in a smaller time frame. Ubuntu terminal version 18.04, Docker, and Tomcat version 7 are required. The command ".run" runs the data center file, and visualization begins. Figure 5.4 depicts a snapshot of the model running.

The data center includes a virtual-based network and some parameters that conduct the blockchain structure. Several tools are used in the evaluation of the model execution results:

- *Big Data Processing:* an open-source java tool that enables the system to process all data without sacrificing throughput as data grows in size. Big

```
*****Datacenter: ClusterChainDatacenter_0*****
User id          Debt
5                5384
6                5384
++++++++++++++++++++++++++++++++++++
*****Datacenter: ClusterChainDatacenter_1*****
User id          Debt
++++++++++++++++++++++++++++++++++++
/////////////////// Map_Reduce Tasks LoadedHost ///////////////////////////////////////////////
UnderLoadedHost Name is :  Host_0
/////////////////// MigrationVm /////////////////////////////////////////
MigrationVm Name is :  VM_1
Simulation completed.

---------- Networks QoS |UBest|XBest|X|FU| ----------
DataC_ID    FUN     ClusterID    Migration    Time    Responce    SimTime
    4       SUCCESS     3             4                 258.06      0.1        258.16
    9       SUCCESS     3             4                 258.06      0.1        258.16
    3       SUCCESS     3             3                 271.18      0.1        271.28
    8       SUCCESS     3             3                 271.18      0.1        271.28
    2       SUCCESS     3             2                 285.71      0.1        285.81
    7       SUCCESS     3             2                 285.71      0.1        285.81
    1       SUCCESS     3             1                 301.88      0.1        301.98
    6       SUCCESS     3             1                 301.88      0.1        301.98
    0       SUCCESS     3             0                 320         0.1        320.1
    5       SUCCESS     3             0                 320         0.1        320.1
---------------------------------------------------------------------------------------------

Fast Data Processing finished!
BUILD SUCCESSFUL (total time: 0 seconds)
```

FIGURE 5.4 Snapshot of "run" model.

data processing provides scalability for allocating more tasks at a given time. It is used to define the entities in the network.

- *CloudSim Simulator Version 3.3:* a java-based simulation toolkit that supports the functionalities of queueing and processing of events. The toolkit enables modeling, simulation, and experimentation in cloud computing and application services. It is used in this model to import algorithms required for the design of cloud data centers. Since the scale of real testbeds, utilization is limited to experiment; simulation tools are a suitable alternative that enables the option to evaluate the hypothesis in an environment that can reproduce tests of cloud computing free of cost and tune the performance blockages before deploying on real clouds. MultiCloudSim library is imported for task scheduling and queuing.
- *Epigenomics Workflow:* used to automate the genome sequencing process associated with resource-intensive tasks. It is a data processing pipeline to automate the execution of various sequence operations, filter out the noisy and contaminating sequences, and map sequences into correct locations (Bharathi et al., 2008).
- *NetBeans (Version 8) Environment:* an open-sourced Integrated Development Environment (IDE) supports the development of a set of modules of Java application types that are used as a browser for input data, computing the cloud networks, and Internet of Things (IoT). NetBeans is run in the Ubuntu terminal.

- *SimJava:* is a discrete-event, process-oriented simulation package. It is an API that augments Java with building blocks for defining and running simulations. The original SimJava was based on HASE++, a C++ simulation library. HASE++ was, in turn, based on Jade's SIM++. SimJava comes with some animation facilities.

The input datasets are through Docker by Cloudlets and brokers that are applied as limited workflow in the network in the format of MIPS. For each of these Cloudlets, data centers are generated with different central processing unit (CPU) cores. The model uses the normal distribution to consider random input data of the virtual machines in control of physical machines. In this model, one must have several servers with different cores, aggregation, and accessibility. A CPU 7 is required to develop the configurations to run the model. The minimum virtual number request (VNR) is considered. At first, the VNR of 0.0001 for the simulation run is applied. The topology is then conducted. Table 5.1 depicts additional simulation specifications.

Next step, data for scheduling, DVFS-based algorithms, and probability models are executed in the NetBeans environment. The total computing capacity is based on the MIPS value. The MIPS value is the basis for building the blockchain structure and is responsible for simulating the developed model. Depending on the complexity of the model, this step can take several hours. However, when the simulation is complete, the results are automatically displayed. The simulation results ".tr" can be indexed and represented with the Chrome web browser. Data sets of BigData, CloudSim, and Bchaincloudsim are added to the platform to analyze the results. The logs are registered on the server. In every virtual machine, different information must be stored based on the userId and the IdShift, which consider the blockchain data on the servers (AliPour, 2021).

Next, the network is developed along with all servers. With a search based on the hostList, it returns the score of servers that DVFS is added on. A list of information is returned on the Cloudlet. The "broker" provides the output in the timeslots and time intervals on the virtual machine. The other parameter, a list based on the proposed fit function method, is an approximate probability of the entire data centers intended to run the blockchain. In the next part, the load

TABLE 5.1

Additional Specifications for the Simulation

No.	Type	MIPS	RAM	Bandwidth	Virtual Machine Monitor	Virtual Machine Capacity	Processing Elements
1	A	300	512–2 GMB	10–100 Mb/s	Xen-IntT	2000–2200 MB	1–3
2	B	500	256–512 GMB	10–100 Mb/s	Xen-IntT	2000–2200 MB	1–3
3	A	300	512–2 GMB	10–100 Mb/s	Xen-IntT	2000–2200 MB	1–3
4	B	500	256–512 GMB	10–100 Mb/s	Xen-IntT	2000–2200 MB	1–3
5	A	300	512–2 GMB	10–100 Mb/s	Xen-IntT	2000–2200 MB	1–3
6	B	500	256–512 GMB	10–100 Mb/s	Xen-IntT	2000–2200 MB	1–3

process is reviewed, and determine when the migration algorithm is needed to be applied. The next part is another Cloudlet that is sent to the hostList and the information that is asked to be returned is energy consumption for each virtual machine scheduling task. Then, the changes in CPU utilization, memory, and the network structure are analyzed when migration is applied to the virtual machine. The next part includes the MIPS results based on DVFS, the quality of the network services, and the simulation runs on both algorithms of the reference model and the proposed model. When this part is completed, it reads the outputs from the brokers. It also builds the chains, tries to connect to each virtual machine (1 to 4), and extracts data related to Cloudlets. The outputs are returned for broker "0" and broker "n-1" and print the ones that receive tasks (AliPour, 2021). The virtual machine parameters include Image size (1000 MB), Ram VR Memory (512), MIPS (250), Bandwidth (1000), number of CPUs (2), and Virtual Machine Manager. The Virtual Machine Manager enables virtualizing multiple operating systems that run concurrently on the host computer.

5.4   SIMULATIVE ANALYSIS

The analysis presents the impact of blockchain technology on disaster supply management resilience using simulation. Several cases are considered to have a multi-facet investigation. The degree of effectiveness of the proposed model is analyzed by varying the input parameters. Hence, the simulation experiments can state the impacts of the proposed model on disaster supply management resilience. Specifically, two research questions are considered:

- Research Question (RQ1): How can the blockchain-enabled model enhance the disaster supply network performance?
- Research Question (RQ2): How can the blockchain-enabled model enhance the disaster supply network resilience?

The accompanying hypotheses are listed as follows:

- RQ Hypothesis (RQH1A): The level of real-time information sharing in the platform is positively related to the response time.
- RQ Hypothesis (RQH1B): The level of real-time information sharing in the platform is positively related to delay in demand provision.
- RQ Hypothesis (RQH2A): The blockchain-enabled model has a positive impact on disaster complexity management.
- RQ Hypothesis (RQH2B): The blockchain-enabled model has a positive impact on monitoring the disaster supply network.

Table 5.2 provides components evaluated for the study to support the research questions and the hypothesis. The simulation conditions are captured as numbers such that 1 is the "Smart contracts-enabled simulation for hyperconnected logistics" (SCESHL) model, 2 is DVFS algorithm applied to the SCESHL model,

TABLE 5.2

Specific Output Metric Components and Their Relation to the Research Topics

Outputs	Description	Research Topic	Related Hypothesis
System Delays	Required time for the tasks to be handled by the blockchain method (Min)	Performance Resilience	• RQH1A • RQH2A
Energy Consumption	Required energy amount to receive the needs requests and process them in the network (kWh)	Resilience	•
System Errors	Unsuccessful migration of tasks handled by virtual machines	Resilience	• RQH2A
Number of Actions	Number of actions that can be supported in every time step of the model (Min)	Performance Resilience	• RQH1B • RQH2B
Successful Migrations	Transfer tasks from the overloaded node to the next block in the virtual machine with the capacity to handle the request	Performance Resilience	• RQH1B • RQH2A • RQH2B
System Throughput	Quantity of data that can be transferred from one location to another within a specific time (MIPS)	Performance	• RQH1A • RQH1B

3 is the proposed blockchain-enabled DVFS-based model, and 4 is the proposed blockchain-enabled DVFS-based model integrated with MAS.

The results of this study are comparable to previously published research (Betti et al., 2020). By comparing present research to previously published work, we can establish credibility and contribute to the current body of knowledge. Moreover, such an approach provides means for verification of the model. It also enables one to compare in terms of improvement—worse, same, or better.

5.4.1 SYSTEM DELAYS

The X-axis shows the number of actions (events) in terms of the functions that a blockchain is handling. The Y-axis shows the delay time in minutes. Figure 5.5 indicates that the proposed model has fewer delays, rooted in fewer errors. Tasks entering the machines within this model are handled at a faster pace. Condition 4, where the relationship between the blockchains and the DVFS is managed using ABM, has fewer errors and delays. Therefore, the proposed model manages delays more efficiently. Moreover, there is a 5% improvement from conditions 1 to 2, from 2 to 3, and from 3 to 4.

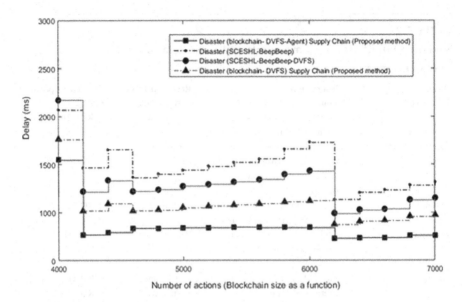

FIGURE 5.5 Diagramming system.

5.4.2 ENERGY CONSUMPTION

Energy consumption is the amount of energy (power) used from the beginning of the task request to the completion of response actions. Figure 5.6 depicts the amount of energy required to receive requests (actions) via the managing tasks in the network until the completion of disaster response actions. The Y-axis represents the power consumption in kWh, while the X-axis is the time in minutes. As the algorithm is executed, at around 30 minutes, the model is optimized around 9.12% compared to the referenced model. Notice too that the proposed model observes the green computing policies.

Conditions 1, 2, 3, and 4 have the same results at the initial stage, which validated the network configurations. As the algorithms are implemented, the proposed model consumes less power than the referenced model. Moreover, Figure 5.6 proves the data centers are consuming less power to provide better results to support the disaster response network despite time complexity, space complexity, and computations. These results suggest that the developed model has less computational complexity than the referenced model.

5.4.3 SYSTEM ERRORS

The y-axis is the error line based on the number of actions and the function producing those actions. The x-axis is the probability of failures. Figure 5.7 depicts all four conditions of the models. Errors are higher in conditions 1 and 2. The blockchain-based processing line includes a high level of actions and complex timing. When the DVFS is applied, the error is improved by almost 8% compared to condition 2. There is a 15% improvement in errors in condition 3 and a 17% improvement in condition 4.

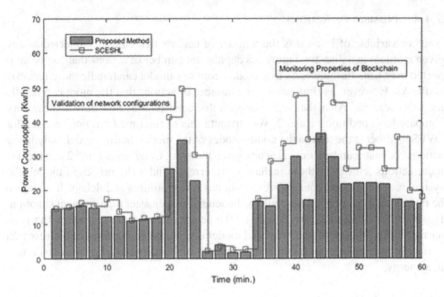

FIGURE 5.6   Diagramming power consumption.

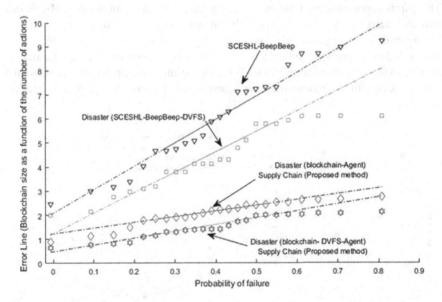

FIGURE 5.7   Diagramming system error line.

Moreover, we learn that when the option of migration is available, the tasks (the requirements requests from the affected population) can be handled more effectively, and the disaster is managed more efficiently. As the errors in the system decrease, the delay time is also reduced. Needless to say, there are more failures in conditions 1 and 2. Therefore, the proposed model is an efficient model for reducing the response time in disaster response.

5.4.4 Number of Actions

Another variable of interest is the number of actions that can be supported at any given moment in a disaster. Figure 5.8 depicts the number of actions that can be supported over time (minutes). As expected, a complex model can handle more requests and tasks. However, as time passes, the number of actions that the more can handle reduces over time. Also, the model suggests that conditions 3 and 4 have less capacity compared to conditions 1 and 2. We attribute this to the time (duration) needed for DVFS to allocate the task to the middle nodes of the model. In this model, only fewer actions/requests can be received at any given moment. Conditions 1 and 2 can receive more actions. However, these include more errors. Under the proposed model, the system receives fewer requests/actions but has fewer failures and delays. It may also be that receiving more actions could be beneficial in a disaster management situation. However, handling tasks aright may be the priority if one is concerned with disaster resilience. Therefore, the proposed model is efficient and optimizes the research variables by having fewer failures and delays in the entire network while consuming less energy.

5.4.5 Successful Migrations

The x-axis represents the total of events that are increasing over time. Migrations indicate data management within all virtual machines and are beneficial in case of having loads of data in the system, such as in the case of disaster situations. Figure 5.9 indicates that the proposed model has 20%–25% more migrations. These adjustments are due to the implementation of the DVFS algorithm in the proposed model. As the rate of successful migrations increases, there would be less damage to the supplies,

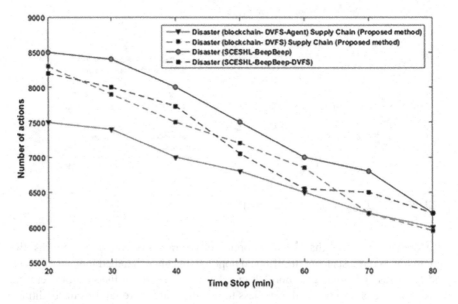

FIGURE 5.8   Diagramming the number of actions.

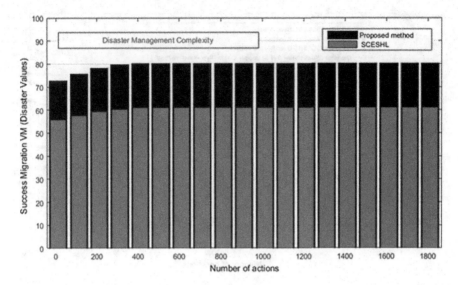

FIGURE 5.9   Diagramming the number of successful migration.

and the outputs have better performance. This figure also addressed disaster management complexity. When the number of events increases, the complexity of disaster management also heightens. The system requires a more successful migration of tasks to manage all the input data effectively. The responsibilities are the inputs of the sensor data, which is considered the stream of input data in the model.

5.4.6   SYSTEM THROUGHPUT

The y-axis is each property's throughput, and the x-axis is the number of input tasks consumed per second. Figure 5.10 depicts the results for four conditions, with conditions 3 and 4 most likely the same as conditions 1 and 2. Moreover, when the DVFS is applied, the system is not changed even though the iteration number gets slightly better. In this case, it is a result of spatial complexity. In condition 4 (i.e., agents have been considered within the DVFS-based model), the throughputs are higher than in condition 1. This depicts that the proposed model is efficient and can be applied as a decision support system for disaster supply chain management.

The time complexity of the disaster-aids network is an exciting aspect of the present research. Interestingly, the proposed model suggests improvement. This improvement is also verifiable by the throughput management. The model can achieve optimal solutions using the DVFS-based model considering blockchain agent-based conditions.

CONCLUSIONS

An iteration is one recalculation of a model during a simulation, while a single simulation consists of many iterations. All uncertain variables are sampled once during each iteration according to their probability distributions. This chapter presents

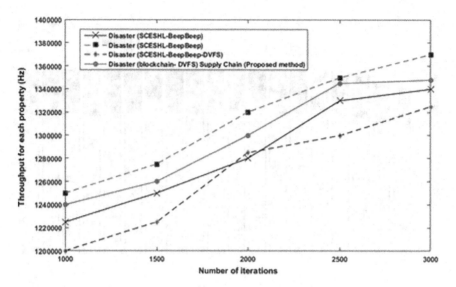

FIGURE 5.10   Diagramming system throughput.

the results of 100 iterations simulating the DVFS model using a data stream. The proposed algorithm uses measures suggested as critical requirements in disaster response. Moreover, we considered the use of simulation to enable repeatability as well as traceability. Under the simulation, under four different conditions (i.e., SCESHL model, DVFS algorithm applied to the SCESHL model, proposed blockchain-enabled DVFS-based model, and proposed blockchain-enabled DVFS-based model integrated with MAS), improvements in the disaster supply chain network are illustrated along considering six outputs: energy consumption, number of actions, successful migrations, system delays, system errors, and system throughput. A case for additional output measures can always be made proposed a blockchain-enabled DVFS-based model integrated with MAS. At the same time, even a single improvement can make a difference in the case of disaster-stricken areas. The implication of this simulation is discussed in the ensuing chapter.

REFERENCES

AliPour, F. S. (2021). *Application of a blockchain enabled model in disaster aids supply network resilience* [Dissertation, Old Dominion University Libraries]. https://doi.org/10.25777/FKR7-A212

Betti, Q., Montreuil, B., Khoury, R., & Hallé, S. (2020). Smart contracts-enabled simulation for hyperconnected logistics. In M. A. Khan, M. T. Quasim, F. Algarni, & A. Alharthi (Eds.), *Decentralised Internet of Things: A blockchain perspective* (pp. 109–149). Springer International Publishing. https://doi.org/10.1007/978-3-030-38677-1_6

Bharathi, S., Chervenak, A., Deelman, E., Mehta, G., Su, M.-H., & Vahi, K. (2008). Characterization of scientific workflows. *2008 Third Workshop on Workflows in Support of Large-Scale Science*, 1–10. https://doi.org/10.1109/WORKS.2008.4723958

Iglesias, C. A., Favenza, A., & Carrera, Á. (2020). A big data reference architecture for emergency management. *Information*, *11*(12), 569. https://doi.org/10.3390/info11120569

Liberatore, F., Pizarro, C., de Blas, C. S., Ortuño, M. T., & Vitoriano, B. (2013). Uncertainty in humanitarian logistics for disaster management: A review. In B. Vitoriano, J. Montero, & D. Ruan (Eds.), *Decision aid models for disaster management and emergencies* (pp. 45–74). Atlantis Press. https://doi.org/10.2991/978-94-91216-74-9_3

6 Implications of the Blockchain-Enabled Model

6.1 VERIFICATION AND VALIDATION

Verification and validation (also abbreviated as V&V) are independent procedures used to check that, for example, a model (or product, service, or system) meets requirements and specifications and fulfills its intended purpose (Hojo, 2004). V&V is a critical aspect quality management system. However, V&V is not created equal across all systems, domains, and problems. For example, V&V in a disaster situation cannot be expected to be the same in less critical situations. Moreover, validation can take on different forms, including prospective, retrospective, full-scale, partial, cross-validation, re-validation/locational (periodical validation), or concurrent validation.

In the present work, we suggest thinking in terms of the contrast between ("in vivo" and "in vitro." *In vivo* is Latin for "within the living" and is the study model for a process or procedure conducted "on" a living system or situation—as in a live disaster. *In vitro* is Latin for "in the glass" and is a study model for a procedure conducted in a controlled environment such as a laboratory—as in a simulation. There are distinctive reasons why each model of study is necessary.

Obviously, each approach offers advantages and disadvantages (Uttekar & Allarakha, 2021), as depicted in Table 6.1.

From Table 6.1, it is easy to see why applying the "in vivo" approach in a disaster situation would be highly irresponsible and even criminal. Nevertheless, the "in vitro" approach, similar to "in vivo" can provide insights for effectively dealing with disaster supply chain and logistic issues—from identifying threats to developing mitigation strategies.

Moreover, the present research contributes to the disaster management body of knowledge. For example, the research can be evaluated against present knowledge. If software package tools are used, the flow of knowledge and different paths of research approaches can be detected. Also, the authors-keyword search analysis provides a landscape that can be used to understand and extend blockchain usability in supply chain management and beyond. In fact, there is an increasing trend toward developing and using models to enhance relief decision-making (Sargent, 2010). Model verification checks the computer program's correctness and computerized model implementation (Sargent, 2010). Models can also be validated to remarkably astonishing satisfactory ranges of accuracy (Sargent, 2010). Although validation can be quite costly, specification verification ensures that the computer software design

TABLE 6.1

Differentiating between In Vivo and In Vitro for Validation

Element of concern	In Vivo	In Vitro
Cost and preparation	• High priced because it involves live experimentations. • Moral dilemmas arise if life, humans or otherwise, is involved.	• Low priced and efficient since life is often not in harm's way. Labs can use animal cells while simulations can use inanimate subjects.
Time	• It can be time-consuming to arrive at the result.	• The results are fairly achieved in a short amount of time.
Results	• Results tend to be very specific, detailed, and accurate.	• Findings may not reflect the realities of live situations; multiple iterations are required for confirmation.
Testing Regulations	• Tend to be strict due to implications on subjects.	• There are more resting regulations.
Applications	• Develop new treatments and research protocols.	• Develop new treatments and research protocols.

and implementation of the conceptual model are reasonable. Implementation verification ensures that the simulated model is implemented according to the requirement. Operational validation determines whether the output behavior of the model has adequate accuracy over the domain of intended pertinence.

A simple approach to verify and validate a model is comparison—comparing the model to other already existing models. In fact, it is standard practice to validate different simulation results through comparative analysis (Sargent, 2010). The present research compares the proposed model to the model proposed by Betti et al. (2020). The model results are the basis for the proposed integrated blockchain DVFS-based model that can enhance performance in disaster supply chain and logistics networks. Moreover, the results suggest a positive relationship between response time and demand requirement with increased and transparent data sharing. The results also suggest that the proposed model enhances resilience in disaster supply chain and logistics, indicated by a model's monitoring and complexity.

Literature's trend toward examining the role of emerging technologies is here to say. In fact, many of these innovations and breakthroughs happening now will transform business and society (Marr, 2022):

- *Computing Power:* Computing power will continue to explode in 2022 and beyond. We now have considerably better cloud infrastructure, and many businesses are re-platforming to the cloud, including a push for better networks such as 5G and 6G to power phones, cars, and wearable devices.
- *Smarter Devices:* Growing computer power enables the creation of smarter devices—including intelligent televisions and autonomous cars with more intelligent robots working alongside humans to complete more tasks.

- *Quantum Computing:* The trend of quantum computing—the processing of information that is represented by special quantum states—enables machines to handle information in a fundamentally different way from traditional computers. Quantum computing will potentially provide computing power that is a trillion times more powerful than what we get from today's advanced supercomputers.
- *Datafication:* Data is a key enabler for many technological trends. Ubiquitous digitization today has led to the availability of enormous amounts of data, with data becoming the key to many aspects of business operations. Data is (and will remain) a cornerstone of understanding business, researching key trends, and insight generation.
- *Artificial Intelligence/Machine Learning:* Organizations and researchers are now using all their data and computing power to provide advanced AI capabilities, including machine vision and language processing.
- *Extended Reality:* Our devices now have more augmented reality (AR) capabilities. However, there is an unprecedented push towards virtual reality (VR), where scenes (and objects) appear real, immersing the user in their surroundings with incredible experiences in the metaverse.
- *Digital Trust*: Blockchain technology, distributed ledgers, and non-fungible tokens (NFTs) are transforming our world, and we will continue to see advances in this technology in 2022 and beyond.
- *3D Printing:* In 2022 and beyond, 3D printing will be the basis for transformation in manufacturing and beyond, from mass-produced customized pieces to concrete for houses, printed food, metal, and composite materials.
- *Genomics:* The 2020 Nobel Prize in Chemistry was awarded to two scientists, Emmanuelle Charpentier and Jennifer A. Doudna, for their work developing a method for genome editing. Genomics, gene editing, and synthetic biology will affect other domains, including crop modification and other medical and biological breakthroughs.
- *New Energy Solutions:* As the world is confronted with the realities of climate change (and global warming), there will be a continued search for advances in the energy to power cars, ships, planes, and trains, including innovations in nuclear power and green hydrogen.

Many of these technologies can be examined for their applications in managing disasters and enhancing resilience. It is not a far leap given that the present research suggests using advanced technology products (e.g., IoT, Bigdata) to provide real-time communication and heighten cooperation. Cooperation and trust among stakeholders (i.e., participants) are especially relevant in disaster situations. Moreover, the use of citation network clusters and co-occurrence of keyword search analysis aided in obtaining a holistic view of state of the art for blockchain applications in disaster supply chain and logistics management. The results of the literature analysis suggest that there is a significant trend toward blockchain adoption in the domain of disaster supply chain and logistics management.

6.2 EMBEDDING RESEARCH IN DISASTER MANAGEMENT CHALLENGES

First, the study results are coherent with previously published peer-reviewed research results. It is already established that disaster supply operations can be enhanced by using real-time information. Moreover, simulation results show that applying blockchain technology and IoT can improve the response process and enhance disaster management resilience. Furthermore, this study focused on the complex aspects of the disaster supply chain, yet the application of the proposed model suggests that it is possible to improve complexity management.

Energy consumption is a crucial limitation of blockchain technology applications. However, with the development and application of the DVFS-based algorithm, the challenge of high energy consumption is reduced. Betti et al. (2020) research is used as a baseline. The simulation results of the proposed model were consistent with the referenced model. Moreover, improvements are also made beyond Betti et al. (2020) research, making the model more effective in tackling the disaster supply chain challenges. A summary of findings in comparison to the present literature is articulated:

- *Finding 1:* in the proposed model application, a positive influence on response time is obtained. This finding is consistent with previous findings that suggest decentralized, reliable collaboration and information sharing can be used as a basis for disaster supply resilience. Therefore, the results confirm that blockchain technology can substantially enhance disaster management (Aranda et al., 2019; Balcik et al., 2010; Dubey et al., 2020; Tatham & Kovács, 2010; Van Wassenhove, 2006).
- *Finding 2:* the model enables collaboration via smart contracts and distributed networks—collaboration is essential for enhanced performance in disaster situations. Therefore, the use of blockchain technology positively impacts real-time event identification (Aranda et al., 2019; L'Hermitte & Nair, 2021; Papadopoulos et al., 2017). Moreover, the results strongly suggest that response depends on the inherent processes instituted for collaboration (Tatham & Kovács, 2010; Van Wassenhove, 2006).
- *Finding 3:* the proposed model hinges on response time as well as demand. The developed simulation clearly suggests that more behaviors can be supported within each timestep—the model can be made more complex with the right resources. Therefore, it is possible to have minimal delays in response time as well as minimal errors (Betti et al., 2020; Lohmer et al., 2020).
- *Finding 4:* the response time is substantial with high process productivity and less failure. Additionally, the model indicates that network information sharing can be considerably shortened. The shortening is attributed to the blockchain-based model's ability to (i) enhance collaboration of the relevant relief performers, (ii) improve the performance of humanitarian logistics, (iii) share real-time data, and (iv) increase trust (Dubey et al., 2020; Khan et al., 2021; L'Hermitte & Nair, 2021).

6.3   MODEL IMPLICATIONS

Suffice it to say that the aim of developing a blockchain-based model with the potential to affect the disaster supply chain and logistics management has been achieved. Furthermore, the research includes measurable simulation experimentation to support the research questions and investigate indications of the functioning mechanisms of the disaster management complexities in responding to the significant number of requirements in disaster response. In this research, these requirements are limited to energy conception, number of actions, successful migration, system delay, system errors, and system throughput. While there remains room for improved requirements, the results indicate substantial progress in disaster supply management attributed to blockchain technology and the DVFS algorithm.

In disaster management, especially those involving the loss of life, one can never have too many tools. The proposed model should be added to a set of tools supporting decision-makers, enabling surveillance and intervention teams during disaster response, and expediting responses to enable speedy recovery. Moreover, we suggest that this model can be used to train those involved in disasters, and the lessons can be used to improve the responses to emergencies (and disaster aid) and the model.

6.4   RESEARCH LIMITATIONS

This study presents the blockchain application research path in disaster aid supply chain and logistics management. Perhaps the present research is only a brick in the wall for those interested in managing disasters using emerging technologies. There remains a need to clarify how the model fits within the larger context of disaster supply chain and logistics management to more effectively address lead time reduction (before, during, and after the disaster), leading to more resilient aid supply networks. To this end, we suggest the following (AliPour, 2021):

- *Resilience Indicators:* Resilience is not a new concept (Gheorghe & Katina, 2014; Holling, 1973). However, there is a need to develop indicators for resilience quantification. These can be the basis for additional experimental investigations on real disaster supply chains and logistics. Disaster supply chains and logistics networks have unique characteristics beyond original ecological systems, including racial biases (Katina, 2016). Moreover, the present study uses data collected from literature; this single since is not representative of all possible scientific contributions to the field of disaster management.
- *Practical Tools*: A learning curve is associated with the developed model, and a novice researcher might not find the model practical. Therefore, we suggest the development of practical mechanisms. Moreover, intelligent platforms can be designed to classify potential hazards and risks, and governance strategies could be re-assessed. Another approach is to clarify the efficacy of the present studies with established cases. For example, emerging technologies could be infused into disaster management systems to enhance some aspects of supply chain new crises.

- *Case Applications*: There is a need for real case applications of the proposed model. Such applications can only serve as the basis for improved models as well as challenge blockchain technology capabilities—a win-win scenario. The adoption of blockchain technology, while currently a challenge, maybe lessened with the popularity of other technological improvements (e.g., IoT, 5G/6G). Therefore, technical challenges of blockchain technology, including computational power distribution, user integrity, confidentiality, the safety of encryption algorithms, and scalability, can also be enhanced as they enhance the model in use.

Notwithstanding, there remain constraints to blockchain technology. First, the authors suggest the obvious challenge: In this research, it is impossible to have a "concrete" validation of the model since it is not applied to a live disaster. However, the lessons learned cannot be ignored, especially compared to previous studies. Moreover, disasters are not going away anytime soon. Using the model and upkeep for running the full version can be costly. There is also latency in the transaction confirmation process, especially for a disaster case spanning a large area with an interconnected system of systems. In the latter, the consensus mechanism will require the entire association to perform complex algorithms for mining, a not-so-easy task (Rajan, 2018). The model needs to address confidentiality aspects in data collection, especially in cases where information must be shared among collaborating agencies during a disaster. Finally, the provided model focuses on a typical concept of a disaster situation. It does not focus on different disaster scenarios where the dynamics and particulars of each disaster are relevant. Since each disaster is unique, there might be a need to develop and implement unique models for each disaster.

Clearly, the issue of integrating blockchain technology into the disaster supply chain and logistics management for resilience is not fully understood. As already suggested, VUCA describes the constant, unpredictable change situations that now form the "new norm" for many leaders in many problem domains, including disaster management. In an attempt to address these issues, a research agenda is needed.

CONCLUSIONS

The use of technology to address problems is not a new idea. However, combining blockchain technology and resilience as the basis for addressing supply chain and logistics challenges in a disaster is somewhat novel. Novel as it may seem, it is not free of challenges. Beyond the technical challenges, see, for example, Arbesman (2016), where there are verification and validation challenges. Verification and validation challenges are associated with independent procedures that can be used to check that a model (product, service, or system) meets requirements and specifications and fulfills its intended purpose (Hojo, 2004). The validation work can take on different forms, including prospective, retrospective, full-scale, partial, cross-validation, re-validation/locational (periodical validation), or concurrent validation. In the present case, we provide retrospective validation since the current model validates an existing model. However, the research also contains aspects of cross-validation in as much as sets of scientific data generated using two or more methods are critically assessed. And while the results suggest that a blockchain DVFS-based model can enhance resilience in disaster supply chain and logistics management networks, there remain challenges associated with developing robust resilience indicators, practical tools, and model case applications. The words "verification" and "validation" are sometimes preceded with "independent" (hence IV&V), indicating that the verification and validation are to be performed by a "disinterested" third party. However, in the case of disaster management, IV&V should neither be performed by a "disinterested" third party nor focus only on traditional V&V. We suggest an "interested" third party and a multi-phase model V&V—at the construction as well as implementation, perhaps as an extension of research in harmony with the research findings.

REFERENCES

AliPour, F. S. (2021). *Application of a blockchain enabled model in disaster aids supply network resilience* [Dissertation, Old Dominion University Libraries]. *https://doi. org/10.25777/FKR7-A212*

Aranda, D. A., Fernández, L. M. M., & Stantchev, V. (2019). Integration of Internet of Things (IoT) and Blockchain to increase humanitarian aid supply chains performance. 2019 5th International Conference on Transportation Information and Safety (ICTIS), 140–145. https://doi.org/10.1109/ICTIS.2019.8883757

Arbesman, S. (2016). *Overcomplicated: Technology at the limits of comprehension.* Current.

Balcik, B., Beamon, B. M., Krejci, C. C., Muramatsu, K. M., & Ramirez, M. (2010). Coordination in humanitarian relief chains: Practices, challenges and opportunities. *International Journal of Production Economics, 126*(1), 22–34. *https://doi.org/10.1016/j. ijpe.2009.09.008*

Betti, Q., Montreuil, B., Khoury, R., & Hallé, S. (2020). Smart contracts-enabled simulation for hyperconnected logistics. In M. A. Khan, M. T. Quasim, F. Algarni, & A. Alharthi (Eds.), *Decentralised internet of things: A blockchain perspective* (pp. 109–149). Springer International Publishing. https://doi.org/10.1007/978-3-030-38677-1_6

Dubey, R., Gunasekaran, A., Bryde, D. J., Dwivedi, Y. K., & Papadopoulos, T. (2020). Blockchain technology for enhancing swift-trust, collaboration and resilience within a humanitarian supply chain setting. *International Journal of Production Research, 58*(11), 3381–3398. https://doi.org/10.1080/00207543.2020.1722860

Gheorghe, A. V., & Katina, P. F. (2014). Editorial: Resiliency and engineering systems—Research trends and challenges. *International Journal of Critical Infrastructures, 10*(3/4), 193–199.

Hojo, T. (2004). *Quality management systems: Process validation guidance* (GHTF/SG3/N99–10:2004). Global Harmonization Task Force. www.imdrf.org/sites/default/files/docs/ghtf/final/sg3/technical-docs/ghtf-sg3-n99-10-2004-qms-process-guidance-04010.pdf

Holling, C. S. (1973). Resilience and stability of ecological systems. *Annual Review of Ecology and Systematics, 4*(1), 1–23. https://doi.org/10.1146/annurev.es.04.110173.000245

Katina, P. F. (2016). Individual and societal risk (RiskIS): Beyond probability and consequence during Hurricane Katrina. In A. J. Masys (Ed.), *Disaster forensics: Understanding root cause and complex causality* (pp. 1–23). Springer International Publishing. https://doi. org/10.1007/978-3-319-41849-0_1

Khan, M., Imtiaz, S., Parvaiz, G. S., Hussain, A., & Bae, J. (2021). Integration of Internet-of-Things with blockchain technology to enhance humanitarian logistics performance. *IEEE Access, 9*, 25422–25436. https://doi.org/10.1109/ACCESS.2021.3054771

L'Hermitte, C., & Nair, N.-K. C. (2021). A blockchain-enabled framework for sharing logistics resources during emergency operations. *Disasters*, *45*(3), 527–554. https://doi.org/10.1111/disa.12436

Lohmer, J., Bugert, N., & Lasch, R. (2020). Analysis of resilience strategies and ripple effect in blockchain-coordinated supply chains: An agent-based simulation study. *International Journal of Production Economics*, *228*, 107882. https://doi.org/10.1016/j.ijpe.2020.107882

Marr, B. (2022, February 21). The top 10 tech trends in 2022 everyone must be ready for now. *Forbes*. www.forbes.com/sites/bernardmarr/2022/02/21/the-top-10-tech-trends-in-2022-everyone-must-be-ready-for-now/

Papadopoulos, T., Gunasekaran, A., Dubey, R., Altay, N., Childe, S. J., & Fosso-Wamba, S. (2017). The role of Big Data in explaining disaster resilience in supply chains for sustainability. *Journal of Cleaner Production*, *142*, 1108–1118. https://doi.org/10.1016/j.jclepro.2016.03.059

Rajan, S. G. (2018). *Analysis and design of systems utilizing blockchain technology to accelerate the humanitarian actions in the event of natural disasters* [Thesis, Massachusetts Institute of Technology]. https://dspace.mit.edu/handle/1721.1/118526

Sargent, R. G. (2010). Verification and validation of simulation models. *Proceedings of the 2010 Winter Simulation Conference*, 166–183. https://doi.org/10.1109/WSC.2010.5679166

Tatham, P., & Kovács, G. (2010). The application of "swift trust" to humanitarian logistics. *International Journal of Production Economics*, *126*(1), 35–45. https://doi.org/10.1016/j.ijpe.2009.10.006

Uttekar, P. S., & Allarakha, S. (2021). Why is in vivo better than in vitro? *MedicineNet*. www.medicinenet.com/why_is_in_vivo_better_than_in_vitro/article.htm

Van Wassenhove, L. N. (2006). Humanitarian aid logistics: Supply chain management in high gear. *Journal of the Operational Research Society*, *57*(5), 475–489. https://doi.org/10.1057/palgrave.jors.2602125.

7 A Research Agenda in Blockchain-Enabled Resilience

7.1 A COROLLARY LESSON

Undoubtedly, humans are living in a moment when they praise themselves with the belief of being well-informed, intelligent, wiser, and capable of making even better judgments—and yet they are incapable of explaining why doomsday fears loom large. It is true that since World War II, remarkable technological advances have affected humanity—agriculture, economy, education, finance, medicine, public health, telecommunications, transportation, and water—and these have affected living standards. At the same time, risks and threats seem to emerge out of thin air, threatening to eradicate species; supervolcanoes, diseases, and nuclear weapons remain constant, and what of the very reliable institutions? Oh well, surely, insurrections are not black swans.

On the one hand, society would like to believe that we are "better" today than we were yesterday, and on the other hand, there are those who believe society is now more complex, fragile, risky, susceptible, uncertain, volatile, and vulnerable, dealing with the previously unprecedented and unimaginable.

We suggest that any attempt to introduce some rationality into this debate should, in principle, be welcomed. And while identification, assessment, analysis, and treatment of risk (and hazards) remain at the heart of disaster management, it would be wise to examine (rigorously) other ideas that can provide enhancement.

In the author's perspective, since humanity is at the center of this debate, for better or worse, the debate must include deontological and epistemological posits, and to this end, the following are suggested, adopted from Gheorghe and Vamanu (1996) when dealing with disaster supply chain and logistics management:

- The likelihood of emergencies in the operation of human enterprises, posing potential risks to the human condition's integrity and material assets/values, can be diminished yet can never be completely ruled out.
- *Corollary I: Low-probability and high-risk crises should be intellectually treated as actual possibilities, and provisions for their mitigation must be timely and adequately ensured.*
    - A disaster crisis and its response should be synergistic factors in sizing the potential (and actual consequences) of abnormalities in the functioning of socio-technical organizations.
- *Corollary II: A mismanaged response to a disaster may be physically more harmful (and morally more ascribable) than an attitude of non-action.*

DOI: 10.1201/9781003336082-7

- *Corollary III: A well-managed disaster must involve and adequately reflect a full assessment of the balance between the possible detriment of the proposed mitigating intervention and the consequences of non-action—the first deontological rule of intervention.*
  - Good emergency planning and preparedness are amenable to the standard requirements of good management involving informed assessment, consequent adequate decision, a functional chain of command, good communications, and efficient response implementation mechanisms.
- *Corollary IV: At the origin of a well-managed disaster is the informed assessment, which can only rely on adequate knowledge*
  - Though disasters can be categorized, no two disasters shall ever be identical. Variability factors feed into disaster typology, including systems, sites, events, space, scale, and time. The variability also includes human behavior, especially under the stresses of a disaster.
- *Corollary V: Disaster managers should always be prepared to operate (i) within wide margins of uncertainty, (ii) under time's pressure, perhaps reacting in near real-time constraints, and (iii) under severe mental strain.*
- *Corollary VI: Disaster managers use various knowledge bases. These knowledge bases must be used in suitable forms under the terms described. Academic knowledge is only one aspect of the knowledge base necessary to deal with a disaster effectively.*
- *Corollary VI: Disaster managers should have all decisions and all conceivable knowledge since the quality of response to a disaster is directly related to the quality of managers' decisions. Objective and subjective factors must be included with the decision support structure.*

Along with the previously mentioned deontological and epistemological postulates, we provide the following commentaries:

- *Commentary I:* While possessing adequate and reliable computer-based decision support (DSS) tends to be the sole operational expression of good disaster preparedness, DSS is only a tool aiding in the decision-making process and should never be allowed to make the final decision. A decision shall remain an irreducible privilege of humans—the second deontological rule of intervention.
- *Commentary II:* While the variability factor evoked would oppose the notion of imposing standards with disaster emergency *identification, assessment, analysis, and treatment* DSS systems, the ubiquitous potential for disasters and their transborder consequences call for standards in disaster management—these standards can take on different forms, including methods, tools, processes, and technologies.

7.2   RESEARCH SYNOPSIS

Thus far, the present authors have illustrated a need for new and innovative approaches to embedding blockchain technology in disaster supply chain and logistics management to enhance the resilience of disaster relief aid networks. In summary, the practice of supply chain and logistics networking, while desirable and capable as is, can benefit from the ideas of resilience, especially in the complexity and ambiguous landscape. Risk, vulnerability, perceptions, and fragility run wild in such situations. And while studies suggest the application of blockchain, there remains a scarcity of literature discussing how blockchain can be utilized to enhance disaster supply chain and logistics network communication: *a gap that the present research has now addressed using a blockchain-enabled DVFS-based model with MAS integration.*

However, at the most fundamental level, rigorous research needs to establish a paradigm for which knowledge claims can be contrasted (Churchman, 1968; Warfield, 1976). Despite this claim, the literature suggests that there isn't one widely accepted approach to knowledge claim (Burrell & Morgan, 1979; Flood & Carson, 1993). As it turns out, this is a discussion related to philosophy and undoubtedly worth exploring, given the relevance of disasters in the 21st century. Moreover, suppose one takes the view of Burrell and Morgan (1979) and extensions of Flood and Carson (1993). In that case, the key issues are ontology, epistemology, methodology, and nature of human beings also relate to knowledge claims. Ontology deals with how an observer views reality. Epistemology deals with how one obtains and communicates knowledge. The nature of humans deals with how they are described concerning environment/systems. Methodology deals with attempts to investigate and obtain knowledge in the world where we find ourselves.

7.2.1 METHODOLOGY

A methodology involves procedures for gaining knowledge about systems and structured processes involved in intervening in and changing systems (Jackson (1991). Following Burrell and Morgan (1979), methodological approaches can be categorized into two opposing extremes: *idiographic* and *nomothetic*. An idiographic view of a methodology supports subjectivity in the research of complex systems, as suggested by Flood and Carson (1993, p. 248):

> The principal concern is to understand the way an individual creates, modifies, and interprets the world. The experiences are seen as unique and particular to the individual rather than general and universal. External reality is questioned. An emphasis is placed on the relativistic nature of the world to such an extent that it may be perceived as not amenable to study using the ground rules of the natural sciences. Understanding can be obtained only by acquiring firsthand knowledge of the subject under investigation.

The opposing view of methodology—nomothetic—supports the traditional scientific method and its reductionist approach to addressing problematic issues (Churchman, 1968, 1971) and is described as follows (Flood & Carson, 1993, pp. 247–248):

> Analyze relationships and regularities between the elements of which the world is composed . . . identification of the elements and the way relationships can be expressed. The methodological issues are concepts themselves, their measurement, and the identification of underlying themes. In essence, there is a search for universal laws that govern the reality that is being observed. Methodologies are based on systematic processes and techniques.

There is no shortage of methodologies to intervene in and change systems. Examples include systems analysis, systems engineering, operations research, complex system governance, critical systems heuristics, interactive planning, organizational cybernetics, organizational learning, sociotechnical systems, soft systems methodology, strategic assumption surfacing and testing, systems dynamics, systems of systems engineering methodology, and total systems intervention). The reader is directed elsewhere to proponents of these methodologies, classifications, descriptions, advantages, and disadvantages (Jackson, 2003, 2019). Each methodology is developed and grounded in certain core conceptual foundations, and the "selection of a method is based in the context of problematic situation and purpose of analysis" (Katina, 2015).

In this regard, a key research question is proposed: what is the methodological basis for blockchain-enabled resilience in disaster supply chain and logistics management? While literature exists that discusses the relationship between blockchain technology and disaster supply chain and logistics management, there is a scarcity of literature discussing methodology in blockchain-enabled resilience for disaster supply chain and logistics management.

In this regard, a key research question is proposed: what is the basis for methodologies for blockchain-enabled resilience in disaster supply chain and logistics management? A need to address pressing issues in the disaster supply chain and logistics is already established, including (and not limited to) resilience, risk, vulnerability, perception, and fragility. And yet, current methodologies, as suggested in the present literature, are insufficient in addressing present challenges—let alone possible future issues. Moreover, present research lends itself towards a nomothetic approach to methodology. However, the disaster management domain includes aspects of idiography, as much as it embraces subjectivity in complex situations, especially when considering elements of risk perception.

7.2.2 EPISTEMOLOGY

An epistemological aspect of research deals with how a researcher (i.e., a system observer) begins understanding problematic situations and communicating knowledge to fellow researchers. This dimension provides the form of knowledge, how knowledge is acquired, and what is considered to be "true" or "false" (Burrell & Morgan, 1979). There are two opposite extremes of epistemology: positivism and anti-positivism. A *positivistic* approach to research indicates that "knowledge is hard, real, and capable of being transmitted in a tangible form" (Flood & Carson, 1993, p. 247). This stance of epistemology supports the idea that it is possible to explain and predict what happens in the social world by searching for regularities and causal relationships between its constituent elements. The growth of knowledge is essentially a cumulative process in which new insights are added to the existing stock of knowledge and false hypotheses eliminated (Burrell & Morgan, 1979). In the anti-positivism view, "knowledge is soft, more subjective, spiritual, or even transcendental—based on experience, insight, and essentially of a personal nature" (Flood & Carson, 1993, p. 247).

In this regard, a key research question is proposed: what is the epistemological basis for blockchain-enabled resilience in disaster supply chain and logistics management? Again, while literature exists that discusses the relationship between blockchain technology and disaster supply chain and logistics management, there is a scarcity of literature discussing epistemology in blockchain-enabled resilience for disaster supply chain and logistics management.

However, the topics of resilience and fragility can be subjective, and as such, present research might be rendered as anti-positivistic. Indeed, this is the case when people hold different views on whether there is a problem with disaster supply chain and logistics and, if they agree, what approach should be taken to address the issue (i.e., is it blockchain technology or something else?). Moreover, the present text also contains aspects of positivism since the presented knowledge is hard and capable of being transmitted in a tangible form—another simulation can be developed.

7.2.3 ONTOLOGY

Ontology deals with the existence of entities and how such entities can be grouped based on similarities and differences. Moreover, ontology can also describe how "an observer views the nature of reality or how concretely the external world might be understood" (Katina et al., 2014, p. 49). Two opposite extremes of ontology are realism and nominalism. Based on Burrell and Morgan (1979) and extrapolations from Flood and Carson (1993), *realism* is captured as "external to the individual imposing itself on individual consciousness; it is a given 'out there' " (p. 247). Realism suggests that reality is objective.

On the other hand, *nominalism* describes reality as a product of individual consciousness. More significantly, nominalism ascribes to the assumption of individual cognition. Under nominalism, "the social world external to individual cognition is made up of nothing more than names, concepts and labels which are used to structure reality" (Burrell and Morgan (1979, p. 4). The utility of "concepts," "labels," and "names" is based on the convenience they offer as tools that can be used to make sense of and describe reality (Flood & Carson, 1993).

In this regard, a key research question is proposed: what is the ontological basis for blockchain-enabled resilience in disaster supply chain and logistics management? While literature exists that discusses the relationship between blockchain technology and disaster supply chain and logistics management, there is a scarcity of literature discussing ontology in blockchain-enabled resilience for disaster supply chain and logistics management.

A case can be made realism is present research such that, for example, the suggested model is the objective. However, a case can also be made for the nominalistic view of the present research, especially in the six outputs (i.e., throughput, successful migrations, delays, errors, number of actions, energy consumption) of the model. The reality is conceived and simulated and the nature, developed, and interpreted based on blockchain technology and disaster supply chain and logistics management. However, in as much as these ideas are partially dependent on the cognition of observers (researchers), one should not be mistaken to assume that, for example, blockchain-enabled resilience is a fallacy of present researchers.

7.2.4 NATURE OF HUMAN BEINGS

The nature of human beings is the final component of research consideration. This aspect provides a stance on humans and their activities in society. It has been suggested that two opposite extremes of *determinism* and *voluntarism* can describe the nature of human beings (Burrell & Morgan, 1979; Flood & Carson, 1993). A *deterministic* view suggests that humans are "mechanistic, determined by situations in the external world; human beings and their experiences are products of their environment; they are conditioned by external circumstances" (Flood & Carson, 1993, p. 247). The *voluntaristic* view suggests that humans are "completely autonomous and free-willed" (Burrell & Morgan, 1979, p. 6) and that, therefore, they have a "creative role [in their environment] and [can] create their environment" (Flood & Carson, 1993, p. 247).

In this regard, a key research question is proposed: what is the human nature underpinnings for blockchain-enabled resilience in disaster supply chain and logistics management? While literature exists that discusses the relationship between blockchain technology and disaster supply chain and logistics management, there is a scarcity of literature discussing the nature of humans in blockchain-enabled resilience for disaster supply chain and logistics management.

At this point in this research, it should be evident that the present authors took human beings as being voluntaristic. They are endowed with the ability to do something regarding disaster issues. They are responsible for developing methodologies, methods, frameworks, models, and techniques that shape research and intervene in disaster situations.

In the end, these philosophical issues elements underpin approaches that can be undertaken in addressing blockchain-enabled resilience for disaster supply chain and logistics management, with enormous implications for potential solutions.

7.3 RESEARCH AGENDA

Attributed to Albert Einstein is the motto that *we cannot solve our problems with the same thinking we used when we created them*. In the present text, we have attempted to offer a viewpoint, perhaps crude, that could be used to address resilience disaster supply chain and logistics management using blockchain technology. Moreover, the results of 100 iterations simulating the DVFS model suggest that levels of improvement in the disaster supply chain network are possible under specific conditions.

Further research is suggested along the stream of resilience indicators, practical tools, and case applications. More importantly, it is only at this point (and considering philosophical underpinnings) that one might realize that there are many more questions than answers—the challenge of creating traction as a theoretically and conceptually grounded approach to improve disaster supply chain and logistics management.

Furthermore, research in blockchain-enabled resilience for disaster supply chain and logistics management is certainly not confined to a prescribed approach or privileged intellectual school of thought. This section provides one additional suggestion to organize development further. These developments are based on previous work on

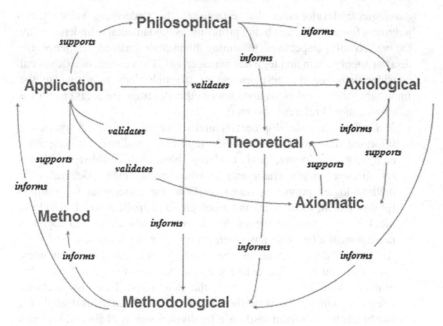

FIGURE 7.1   A balanced view of blockchain-enabled resilience development of disaster supply chain and logistics management.

emerging knowledge (Keating & Katina, 2011; Keating et al., 2022; Keating, 2014). Figure 7.1 provides a framework for the holistic development of blockchain-enabled resilience for disaster supply chain and logistics management.

This framework suggests a purposeful consideration and balanced development along several interrelated lines of inquiry, including philosophical, theoretical, axiological, methodological, axiomatic, method, and application. Even though cogent arguments might be made for one development area having priority over another, what is absolute is that exclusion of any of the areas will not support holistic field development:

- *Philosophical:* development directed at establishing a theoretically consistent articulation of blockchain-enabled resilience for disaster supply chain and logistics management. The emerging system of values and beliefs providing grounding for theoretical development is the primary contribution of this area. A strong, coherent, and articulated philosophical grounding is essential to provide a foundation upon which other field developments can be consistently based.
- *Theoretical:* development focused on explaining system governance phenomena and developing explanatory models and testable conceptual frameworks. The range of theoretical developments advances understanding of the field and the phenomena of central concern. The theoretical development of the field must be actively pursued and not left to chance.

- *Axiological:* development that establishes the underlying value, value judgment frameworks, and belief propositions fundamental to understanding the variety of perspectives informing blockchain-enabled resilience for disaster supply chain and logistics management. The absence of axiological considerations for the development of the field fails to recognize the important value foundations upon which other development areas can utilize as a foundational reference point(s).
  - *Methodological:* development is undertaken to establish theoretically informed frameworks that provide high-level guidance for designing, analyzing, deploying, and evolving blockchain-enabled resilience for disaster supply chain and logistics management. Generalizable methodologies provide a transition from the conceptual foundations (philosophical, theoretical, and axiological) to applications that address blockchain-enabled resilience for disaster supply chain and logistics management along with any emergent issues in the domain.
  - *Axiomatic:* development of the existing and emerging principles, concepts, and laws that define the field and constitute the "taken for granted" knowledge upon which the field rests. This also includes integrating knowledge from other informing and related fields/disciplines. For blockchain-enabled resilience for disaster supply chain and logistics management, the grounding in the axioms and supporting propositions of general systems theory (Klier et al., 2022; von Bertalanffy, 1968) might provide a strong starting point for further axiomatic development.
  - *Method:* development focused on generating the specific models, technologies, standards, processes, and tools for blockchain-enabled resilience for disaster supply chain and logistics management. This is the development of practitioners' supporting tool sets and capabilities. Based on the solid conceptual foundations provided by other field development areas, the methods should be compatible with the field's philosophical, methodological, axiomatic, and axiological predispositions. This approach encourages consistency in the development of methods.
  - *Application:* This emphasizes the advancement of the practice of blockchain-enabled resilience for disaster supply chain and logistics management through the deployment of conceptually sound technologies and methods. Applications not rooted in the conceptual foundations of the field are not likely to be either consistent or conceptually congruent with the deeper underpinnings upon which the field rests. As such, applications void of the field's philosophical, theoretical, and axiomatic foundations are not likely to produce the intended utility for which they have been designed.

Even though cogent arguments might be made for one development area having priority over another, what is absolute is that exclusion of any of the areas will not support holistic field development. Moreover, in this high-level change, specific questions might be crafted based on the needs of stakeholders. For example, the academically inclined might be interested in frameworks that can be constructed (or derived) from existing methodologies (e.g., soft systems methodology and systems of systems engineering) to guide blockchain-enabled resilience for disaster supply chain and logistics management. On the other hand, policy-makers might be interesting in the implications of, for example, voluntarism/determinism before, during, and after a disaster. Simply having a blockchain-enabled resilience model does not mean effective communication among stakeholders will occur in a disaster supply chain and logistics management. These research questions might be addressed at multiple levels, including specific local, regional, and state scales.

CONCLUSIONS

It is indisputable that having new and innovative approaches to address issues in disaster relief aid networks is a good thing. However, using blockchain technology to enhance resilience in disaster supply chain and logistics management is a new kid on the block. And although the simulated application of the developed blockchain DVFS-based model enhances the performance of disaster supply chain and logistics networks, much research remains. It would be "easy" to focus on the limitations of the present research, including the development of resilience indicators, practical tools, and case applications—all worthwhile endeavors. However, rigorous research is the basis for knowledge claims. In this case, a knowledge claim discussion cannot escape philosophical adventures, which tend to include issues of ontology, epistemology, methodology, and the nature of humans. This discussion is the basis for the proposed framework for purposeful and balanced development in blockchain-enabled resilience for disaster supply chain and logistics management. Several interrelated inquiry lines are proposed, including philosophical, theoretical, axiological, methodological, axiomatic, method, and application. It is via the understanding of these deep conceptually and consistent underpinnings that knowledge can be made.

REFERENCES

Burrell, G., & Morgan, G. (1979). *Sociological paradigms and organisational analysis*. Ashgate Publishing.

Churchman, C. W. (1968). *Challenge to reason*. McGraw-Hill.

Churchman, C. W. (1971). *The design of inquiring systems*. Basic Books.

Flood, R. L., & Carson, E. R. (1993). *Dealing with complexity: An introduction to the theory and application of systems science*. Plenum Press.

Gheorghe, A. V., & Vamanu, D. V. (1996). *Emergency planning knowledge*. vdf Hochschulverlag AG an der ETH Zurich.

Jackson, M. C. (1991). *Systems methodology for the management sciences*. Plenum Press.

Jackson, M. C. (2003). *Systems thinking: Creative holism for managers*. John Wiley & Sons Ltd.

Jackson, M. C. (2019). *Critical systems thinking and the management of complexity* (1st ed.). Wiley.

Katina, P. F. (2015). *Systems theory-based construct for identifying metasystem pathologies for complex system governance* [PhD, Old Dominion University].

Katina, P. F., Keating, C. B., & Jaradat, R. M. (2014). System requirements engineering in complex situations. *Requirements Engineering*, *19*(1), 45–62. https://doi.org/10.1007/s00766-012-0157-0

Keating, C. B. (2014). Governance implications for meeting challenges in the system of systems engineering field. *2014 9th International Conference on System of Systems Engineering (SOSE)*, 154–159. https://doi.org/10.1109/SYSOSE.2014.6892480

Keating, C. B., & Katina, P. F. (2011). Systems of systems engineering: Prospects and challenges for the emerging field. *International Journal of System of Systems Engineering*, 2(2/3), 234–256. https://doi.org/10.1504/IJSSE.2011.040556

Keating, C. B., Katina, P. F., Chesterman, C. W., & Pyne, J. C. (Eds.). (2022). *Complex system governance: Theory and practice*. Springer International Publishing. https://link.springer.com/book/10.1007/978-3-030-93852-9

Klier, S. D., Nawrotzki, R. J., Salas-Rodríguez, N., Harten, S., Keating, C. B., & Katina, P. F. (2022). Grounding evaluation capacity development in systems theory. *Evaluation*, 28(2), 231–251. https://doi.org/10.1177/13563890221088871

von Bertalanffy, L. (1968). *General system theory: Foundations, developments, applications*. George Braziller.

Warfield, J. N. (1976). *Societal systems: Planning, policy and complexity*. Wiley-Interscience.

Glossary of Terms

This section contains a glossary of the current knowledge domain terms articulated in the *Developing and Maintaining Emergency Operations Plans* (Fugate, 2010) as suggested by the Federal Emergency Management Agency (FEMA), an agency of the United States Department of Homeland Security. The listed terms have been used in the present text or are closely related to the domain knowledge. In general, explanations of concepts relevant to the present topic are provided. A reader might also use this section to reference material found elsewhere:

ACCESS AND FUNCTIONAL NEEDS

Those actions, services, accommodations, and programmatic, architectural, and communication modifications that a covered entity must undertake or provide to afford individuals with disabilities a full and equal opportunity to use and enjoy programs, services, activities, goods, facilities, privileges, advantages, and accommodations in the most integrated setting. These actions are in light of the exigent circumstances of the emergency and the legal obligation to undertake advance planning and prepare to meet the disability-related needs of individuals who have disabilities as defined by the Americans with Disabilities Act Amendments Act of 2008, P.L. 110–325, and those associated with them.

Access and functional needs may include modifications to programs, policies, procedures, architecture, equipment, services, supplies, and communication methods. Examples of "access and functional needs" services may include a reasonable modification of a policy, practice, or procedure or the provision of auxiliary aids and services to achieve effective communication, including but not limited to the following:

- An exception for service animals in an emergency shelter where there is a no-pets policy
- The provision of way-finding assistance to someone who is blind to orient to new surroundings
- The transferring and provision of toileting assistance to an individual with a mobility disability
- The provision of an interpreter to someone who is deaf and seeks to fill out paperwork for public benefits.

AMERICAN RED CROSS

A nongovernmental humanitarian organization led by volunteers provides relief to disaster victims and helps people prevent, prepare for, respond to, and recover from emergencies. The American Red Cross accomplishes this through services consistent with its Congressional Charter and the Principles of the International Red Cross Movement.

ATTACK

A hostile action taken against the United States by foreign forces or terrorists, resulting in the destruction of or damage to military targets, injury or death to the civilian population or damage to or destruction of public and private property.

CAPABILITIES-BASED PLANNING

Planning, under uncertainty, to provide capabilities suitable for a wide range of threats and hazards while working within an economic framework that necessitates prioritization and choice. Capabilities-based planning addresses uncertainty by analyzing a wide range of scenarios to identify required capabilities.

CHECKLIST

Written (or computerized) enumeration of actions to be taken by an individual or organization meant to aid memory rather than provide detailed instruction.

CITIZEN CORPS

A community-based program administered by FEMA includes Citizen Corps councils and other programs that bring government and nongovernmental entities together to conduct all-hazards emergency preparedness and operations. Through its network of state, territorial, tribal, and local councils, Citizen Corps increases community preparedness and response capabilities via collaborative planning, public education, outreach, training, and volunteer service. Additionally, programs like the Community Emergency Response Team Program train members of the public in basic disaster response skills, such as fire safety, light search and rescue, team organization, and disaster medical operations.

COMMUNITY

Community has more than one definition. Each use depends on the context:

- A political or geographical entity that has the authority to adopt and enforce laws and ordinances for the area under its jurisdiction. In most cases, the community is an incorporated town, city, township, village, or unincorporated area of a county. However, each state defines its own political subdivisions and forms of government.
- A group of individuals (community of interest) who have a religion, a lifestyle, activity interests, an interest in volunteer organizations, or other characteristics in common. These communities may belong to more than one geographic community. Examples include faith-based and social organizations,

nongovernmental and volunteer organizations, private service providers, critical infrastructure operators, and local and regional corporations.

CONSEQUENCE

An effect of an incident or occurrence.

DAM

A barrier built across a watercourse for the purpose of impounding, controlling, or diverting the flow of water.

DAMAGE ASSESSMENT

The process used to appraise or determine the number of injuries and deaths, damage to public and private property, and status of key facilities and services (e.g., hospitals and other health care facilities, fire and police stations, communications networks, water and sanitation systems, utilities, transportation networks) resulting from a human-caused or natural disaster.

DISABILITY

According to the Americans with Disabilities Act, the term "individual with a disability" refers to "a person who has a physical or mental impairment that substantially limits one or more major life activities, a person who has a history or record of such an impairment, or a person who is regarded by others as having such an impairment." The term "disability" has the same meaning as that used in the Americans with Disabilities Act Amendments Act of 2008, P.L. 110–325, as incorporated into the Americans with Disabilities Act. See *www.ada.gov/pubs/ada.htm* for the definition and specific changes to the Americans with Disabilities Act text. State laws and local ordinances may also include individuals outside the federal definition.

DISASTER

An occurrence of a natural catastrophe, technological accident, or human-caused incident that has resulted in severe property damage, deaths, and/or multiple injuries. As used in this guide, a "large-scale disaster" is one that exceeds the response capability of the local jurisdiction and requires state and potentially federal involvement. As used in the Robert T. Stafford Disaster Relief and Emergency Assistance Act (Stafford Act), a "major disaster" is "any natural catastrophe [. . .] or, regardless of cause, any fire, flood, or explosion, in any part of the United States, which in the determination of the President causes damage of sufficient severity and magnitude to warrant major disaster assistance under [the] Act to supplement the efforts and available resources of states, local governments, and disaster relief organizations in alleviating the damage, loss, hardship, or suffering caused thereby" (Stafford Act, Sec. 102(2), 42 U.S.C. 5122(2)).

EARTHQUAKE

The sudden motion or trembling of the ground produced by abrupt displacement of rock masses, usually within the upper 10 to 20 miles of the earth's surface.

EMERGENCY

Any incident, whether natural or human caused, that requires responsive action to protect life or property. Under the Stafford Act, an emergency "means any occasion or instance for which, in the determination of the President, Federal assistance is needed to supplement state and local efforts and capabilities to save lives and to protect property and public health and safety, or to lessen or avert the threat of a catastrophe in any part of the United States" (Stafford Act, Sec. 102(1), 42 U.S.C. 5122(1)).

EMERGENCY ASSISTANCE

According to the National Response Framework, emergency assistance is "[a]ssistance required by individuals, families, and their communities to ensure that immediate needs beyond the scope of the traditional 'mass care' services provided at the local level are addressed. These services include: support to evacuations (including registration and tracking of evacuees); reunification of families; provision of aid and services to special needs populations; evacuation, sheltering, and other emergency services for household pets and services animals; support to specialized shelters; support to medical shelters; nonconventional shelter management; coordination of donated goods and services; and coordination of voluntary agency assistance."

EMERGENCY MEDICAL SERVICES

Services, including personnel, facilities, and equipment, are required to ensure proper medical care for the sick and injured from the time of injury to the time of final disposition (which includes medical disposition within a hospital, temporary medical facility, or special care facility; release from the site; or being declared dead). Further, emergency medical services specifically include those services immediately required to ensure proper medical care and specialized treatment for patients in a hospital and coordination of related hospital services.

EMERGENCY OPERATIONS CENTER

The physical location at which the coordination of information and resources to support incident management (on-scene operations) activities normally takes place. An emergency operations center may be a temporary facility or may be located in a more central or permanently established facility, perhaps at a higher level of organization within a jurisdiction. Emergency operations centers may be organized by major functional disciplines (e.g., fire, law enforcement, medical services), by jurisdiction (e.g., Federal, state, tribal, regional, city, county), or by some combination thereof.

EMERGENCY OPERATIONS PLAN

The ongoing plan is maintained by various jurisdictional levels for responding to a wide variety of potential hazards. It describes how people and property will be protected; details who is responsible for carrying out specific actions; identifies the personnel, equipment, facilities, supplies, and other resources available; and outlines how all actions will be coordinated.

EMERGENCY SUPPORT FUNCTION

Used by the federal government and many state governments as the primary mechanism at the operational level to organize and provide assistance. Emergency support functions align categories of resources and provide strategic objectives for their use. Emergency support functions use standardized resource management concepts such as typing, inventorying, and tracking to facilitate the dispatch, deployment, and recovery of resources before, during, and after an incident.

EVACUATION

The organized, phased, and supervised withdrawal, dispersal, or removal of civilians from dangerous or potentially dangerous areas and their reception and care in safe areas:

- A *spontaneous evacuation* occurs when residents or citizens in the threatened areas observe an incident or receive unofficial word of an actual or perceived threat and, without receiving instructions to do so, elect to evacuate the area. Their movement, means, and direction of travel are unorganized and unsupervised.
- A *voluntary evacuation* is a warning to persons within a designated area that a threat to life and property exists or is likely to exist in the immediate future. Individuals issued this type of warning or order are *not required* to evacuate; however, it would be to their advantage to do so.
- A *mandatory or directed evacuation* is a warning to persons within the designated area that an imminent threat to life and property exists and individuals *must* evacuate in accordance with the instructions of local officials.

EVACUEES

All persons removed or moving from areas threatened or struck by a disaster.

FEDERAL COORDINATING OFFICER

The official appointed by the president to execute Stafford Act authorities, including the commitment of FEMA resources and mission assignments of other federal departments or agencies. In all cases, the federal coordinating officer represents the FEMA administrator in the field to discharge all FEMA responsibilities for the response and recovery efforts underway. For Stafford Act incidents, the federal coordinating officer is the primary federal representative with whom the state coordinating officer and other response officials interface to determine the most urgent needs and to set objectives for an effective response in collaboration with the Unified Coordination Group.

FLOOD

A general and temporary condition of partial or complete inundation of normally dry land areas from overflow of inland or tidal waters, unusual or rapid accumulation or runoff of surface waters, or mudslides/mudflows caused by accumulation of water.

GOVERNOR'S AUTHORIZED REPRESENTATIVE

An individual empowered by a governor to (i) execute all necessary documents for disaster assistance on behalf of the state, including certification of applications for public assistance; (ii) represent the governor of the impacted state in the Unified Coordination Group, when required; (iii) coordinate and supervise the state disaster assistance program to include serving as its grant administrator; and (iv) identify, in coordination with the state coordinating officer, the state's critical information needs for incorporation into a list of essential elements of information.

HAZARD

A natural, technological, or human-caused source or cause of harm or difficulty.

HAZARDOUS MATERIAL

Any substance or material that, when involved in an accident and released in sufficient quantities, poses a risk to people's health, safety, and/or property. These substances and materials include explosives, radioactive materials, flammable liquids or solids, combustible liquids or solids, poisons, oxidizers, toxins, and corrosive materials.

HOUSEHOLD PET

According to FEMA Disaster Assistance Policy 9253.19, "[a] domesticated animal, such as a dog, cat, bird, rabbit, rodent, or turtle, that is traditionally kept in the home for pleasure rather than for commercial purposes, can travel in commercial carriers, and be housed in temporary facilities. Household pets do not include reptiles (except turtles), amphibians, fish, insects/arachnids, farm animals (including horses), and animals kept for racing purposes." This definition is used by FEMA to determine assistance that FEMA will reimburse and is the definition used in the production of this guide. Individual jurisdictions may have different definitions based on other criteria.

HURRICANE

A tropical cyclone, formed in the atmosphere over warm ocean areas, in which wind speeds reach 74 miles per hour or more and blow in a large spiral around a relatively calm center or eye. Circulation is counterclockwise in the Northern Hemisphere and clockwise in the Southern Hemisphere.

INCIDENT

An occurrence or event—natural, technological, or human-caused—that requires a response to protect life, property, or the environment (e.g., major disasters, emergencies, terrorist attacks, terrorist threats, civil unrest, wildland and urban fires, floods, hazardous materials spills, nuclear accidents, aircraft accidents, earthquakes, hurricanes, tornadoes, tropical storms, tsunamis, war-related disasters, public health and medical emergencies, and other occurrences requiring an emergency response).

INCIDENT COMMAND SYSTEM

A standardized on-scene emergency management construct specifically designed to provide an integrated organizational structure that reflects the complexity and demands of single or multiple incidents, without being hindered by jurisdictional boundaries. The Incident Command System is the combination of facilities, equipment, personnel, procedures, and communications operating within a common organizational structure and designed to aid in the management of resources during incidents. It is used for all kinds of emergencies and is applicable to small as well as large and complex, incidents. The Incident Command System is used by various jurisdictions and functional agencies, both public and private, to organize field-level incident management operations.

INCIDENT MANAGEMENT ASSISTANCE TEAM

A national-based or regional-based team composed of SMEs and incident management professionals, usually composed of personnel from multiple federal departments and agencies, which provide incident management support during a major incident.

JOINT FIELD OFFICE

The primary federal incident management field structure. The Joint Field Office is a temporary federal facility that provides a central location for the coordination of federal, state, territorial, tribal, and local governments and private sector and nongovernmental organizations with primary responsibility for response and recovery. The Joint Field Office structure is organized, staffed, and managed in a manner consistent with National Incident Management System principles and is led by the Unified Coordination Group. Although the Joint Field Office uses an Incident Command System structure, the Joint Field Office does not manage on-scene operations.

Instead, the Joint Field Office focuses on providing support to on-scene efforts and conducting broader support operations that may extend beyond the incident site.

JOINT INFORMATION CENTER

A facility established to coordinate all incident-related public information activities. It is the central point of contact for all news media. Public information officials from all participating agencies should co-locate at the Joint Information Center.

JURISDICTION

Jurisdiction has more than one definition. Each use depends on the context:

- *A Range or Sphere of Authority:* public agencies have jurisdiction at an incident related to their legal responsibilities and authority. Jurisdictional authority at an incident can be political or geographical (e.g., city, county, tribal, state, or federal boundary lines) or functional (e.g., law enforcement, public health).
- A political subdivision (e.g., federal, state, county, parish, municipality) with the responsibility for ensuring public safety, health, and welfare within its legal authorities and geographic boundaries.

LIKELIHOOD

Estimate of the potential for an incident's occurrence.

LIMITED ENGLISH PROFICIENCY

Persons who do not speak English as their primary language and who have a limited ability to read, speak, write, or understand English.

MASS CARE

The actions that are taken to protect evacuees and other disaster victims from the effects of the disaster. Activities include mass evacuation, mass sheltering, mass feeding, access and functional needs support, and household pet and service animal coordination.

MITIGATION

Activities providing a critical foundation in the effort to reduce the loss of life and property from natural and/or human-caused disasters by avoiding or lessening the impact of a disaster and providing value to the public by creating safer communities. Mitigation seeks to fix the cycle of disaster damage, reconstruction, and repeated damage. These activities or actions, in most cases, will have a long-term sustained effect.

NATIONAL INCIDENT MANAGEMENT SYSTEM

A set of principles that provides a systematic, proactive approach guiding government agencies at all levels, nongovernmental organizations, and the private sector to work seamlessly to prevent, protect against, respond to, recover from, and mitigate the effects of incidents, regardless of cause, size, location, or complexity, in order to reduce the loss of life or property and harm to the environment.

NATIONAL RESPONSE FRAMEWORK

This document establishes a comprehensive, national, all-hazards approach to domestic incident response. It serves as a guide to enable responders at all levels of government and beyond to provide a unified national response to a disaster. It defines the key principles, roles, and structures that organize the way US jurisdictions plan and respond.

NONGOVERNMENTAL ORGANIZATION

An entity with an association that is based on the interests of its members, individuals, or institutions. It is not created by a government, but it may work cooperatively with government. Such organizations serve a public purpose and are not for private benefit. Examples of nongovernmental organizations include faith-based charity organizations and the American Red Cross.

PLANNING ASSUMPTIONS

Parameters that are expected and used as a context, basis, or requirement for the development of response and recovery plans, processes, and procedures. If a planning assumption is not valid for a specific incident's circumstances, the plan may not be adequate to ensure response success. Alternative methods may be needed. For example, if a decontamination capability is based on the planning assumption that the facility is not within the zone of release, this assumption must be verified at the beginning of the response.

PREPAREDNESS

Actions that involve a combination of planning, resources, training, exercising, and organizing to build, sustain, and improve operational capabilities. Preparedness is the process of identifying the personnel, training, and equipment needed for a wide range of potential incidents and developing jurisdiction-specific plans for delivering capabilities when needed for an incident.

PREVENTION

Actions to avoid an incident or to intervene to stop an incident from occurring. Prevention involves actions to protect lives and property. It involves applying intelligence and other information to a range of activities that may include such countermeasures as deterrence operations; heightened inspections; improved surveillance and security operations; investigations to determine the full nature and source of the threat; public health and agricultural surveillance and testing processes; immunizations, isolation, or quarantine; and, as appropriate, specific law enforcement operations aimed at deterring, preempting, interdicting, or disrupting illegal activity and apprehending potential perpetrators and bringing them to justice.

PROTECTED GROUP

A group of people qualified for special protection by a law, policy, or similar authority. For example, Title VI of the Civil Rights Act of 1964 protects against discrimination on the grounds of race, color, or national origin.

PROTECTION

Actions to reduce or eliminate a threat to people, property, and the environment. Primarily focused on adversarial incidents, the protection of critical infrastructure and key resources is vital to local jurisdictions, national security, public health and safety, and economic vitality. Protective actions may occur before, during, or after an incident and prevent, minimize, or contain the impact of an incident.

RECOVERY

The development, coordination, and execution of service and site restoration plans; the reconstitution of government operations and services; individual, private sector, nongovernmental, and public assistance programs to provide housing and to promote restoration; long-term care and treatment of affected persons; additional measures for social, political, environmental, and economic restoration; evaluation of the incident to identify lessons learned; post-incident reporting; and development of initiatives to mitigate the effects of future incidents.

RESOURCE MANAGEMENT

A system for identifying available resources at all jurisdictional levels to enable timely, efficient, and unimpeded access to resources needed to prepare for, respond to, or recover from an incident. Resource management under the National Incident Management System includes mutual aid and assistance agreements; the use of special federal, state, territorial, tribal, and local teams; and resource mobilization protocols.

RESPONSE

Immediate actions to save and sustain lives, protect property and the environment, and meet basic human needs. Response also includes the execution of plans and actions to support short-term recovery.

RISK

The potential for an unwanted outcome resulting from an incident or occurrence, as determined by its likelihood and the associated consequences.

RISK ANALYSIS

A systematic examination of the components and characteristics of risk.

RISK ASSESSMENT

A product or process that collects information and assigns values to risks for the purpose of informing priorities, developing or comparing courses of action, and informing decision-making.

RISK IDENTIFICATION

The process of finding, recognizing, and describing potential risks.

RISK MANAGEMENT

The process of identifying, analyzing, assessing, and communicating risk and accepting, avoiding, transferring, or controlling it to an acceptable level at an acceptable cost.

SCENARIO

Hypothetical situation composed of a hazard, an entity impacted by that hazard, and associated conditions including consequences when appropriate.

SCENARIO-BASED PLANNING

A planning approach that uses a hazard vulnerability assessment to assess the hazard's impact on an organization on the basis of various threats that the organization could encounter. These threats (e.g., hurricane, terrorist attack) become the basis of the scenario.

SENIOR OFFICIAL

The elected or appointed official who, by statute, is charged with implementing and administering laws, ordinances, and regulations for a jurisdiction. He or she may be a mayor, city manager, etc.

SERVICE ANIMAL

Any guide dog, signal dog, or other animal individually trained to assist an individual with a disability. Service animals' jobs include but are not limited to the following:

- Guiding individuals with impaired vision
- Alerting individuals with impaired hearing (to intruders or sounds such as a baby's cry, the doorbell, and fire alarms)
- Pulling a wheelchair
- Retrieving dropped items
- Alerting people of impending seizures
- Assisting people who have mobility disabilities with balance or stability

STANDARD OPERATING PROCEDURE/GUIDELINE

A reference document or operations manual that provides the purpose, authorities, duration, and details for the preferred method of performing a single function or a number of interrelated functions in a uniform manner.

STATE COORDINATING OFFICER

The individual appointed by the governor to coordinate state disaster assistance efforts with those of the federal government. The state coordinating officer plays a critical role in managing the state response and recovery operations following Stafford Act declarations. The governor of the affected state appoints the state coordinating officer, and lines of authority flow from the governor to the state coordinating officer, following the state's policies and laws.

STORM SURGE

A dome of seawater created by strong winds and low barometric pressure in a hurricane that causes severe coastal flooding as the hurricane strikes land.

TERRORISM

Activity that involves an act that is dangerous to human life or potentially destructive of critical infrastructure or key resources, is a violation of the criminal laws of the United States or of any state or other subdivision of the United States, and appears to be intended to intimidate or coerce a civilian population, to influence the policy of a government by intimidation or coercion, or to affect the conduct of a government by mass destruction, assassination, or kidnapping.

TORNADO

A local atmospheric storm, generally of short duration, formed by winds rotating at very high speeds, usually in a counterclockwise direction. The vortex, up to several hundred yards wide, is visible to the observer as a whirlpool-like column of winds rotating about a hollow cavity or funnel. Winds can be as low as 65 miles per hour but may reach 300 miles per hour or higher.

TSUNAMI

Sea waves produced by an undersea earthquake. Such sea waves can reach a significant height resulting in damage or devastation to coastal cities and low-lying coastal areas.

UNCERTAINTY

The degree to which a calculated, estimated, or observed value may deviate from the true value.

VULNERABILITY

A physical feature or operational attribute that renders an entity open to exploitation or susceptible to a given hazard.

WARNING

The alerting of emergency response personnel and the public to the threat of extraordinary danger and the related effects that specific hazards may cause.

REFERENCE

Fugate, W. C. (2010). *Developing and maintaining emergency operations plans: Comprehensive preparedness guide (CPG) 101*. Federal Emergency Management Agency. www.ready.gov/sites/default/files/2019-06/comprehensive_preparedness_guide_developing_and_maintaining_emergency_operations_plans.pdf

Index

Note: Page numbers in *italics* indicates figures and page numbers in **bold** indicates tables on the corresponding page.

Printed in the United States
by Baker & Taylor Publisher Services